T0301414

VARIATION RISK MANAGEMENT

VARIATION RISK MANAGEMENT

Focusing Quality Improvements in Product Development and Production

Anna C. Thornton

WILEY

JOHN WILEY & SONS, INC.

Published by John Wiley & Sons, Inc., Hoboken, New Jersey
Published simultaneously in Canada

For general information on our other products and services or for technical support, please contact our
Customer Care Department within the United States at (800) 762-2974, outside the United States at
(317) 572-3993 or fax (317) 572-4002.

Wiley also publishes its books in a variety of electronic formats. Some content that appears in print may
not be available in electronic books. For more information about Wiley products, visit our web site at
www.wiley.com.

Library of Congress Cataloging-in-Publication Data:

Thornton, Anna C., 1968–
    Variation risk management : focusing quality improvements in product
development and production / Anna C. Thornton.
        p.   cm.
    ISBN 0-471-44679-3   (Cloth)
    1. Quality control.   2.   Production engineering.   3.   Risk management.
I.   Title.
    TS156.T46   2004
    658.5'62—dc22

                                                            2003017776

10   9   8   7   6   5   4   3   2   1

To my husband, David, and daughter Alexandra

CONTENTS

PREFACE

As a new assistant professor at MIT, I was searching around for a research focus when I stumbled across the term "Key Characteristic" during work on a project on Fast and Flexible Communication of Engineering Information in the Aerospace Industry. Intrigued, I called contacts at Boeing, Kodak, Ford, and other large manufacturing organizations. Everyone I spoke with described the need to identify key characteristics (KCs)—that is, those parameters in manufacturing that, when they vary, impact a product's final performance. However, no one had a comprehensive definition of KCs, a consistent process for identifying them, or a standard method for responding to them. Key characteristic–based processes in use were varied and often misinterpreted; this situation was exacerbated by the fact that every company had multiple definitions of KCs and left each team to define its own process for identifying and reacting to KCs.

At this time, tools such as assembly variation modeling software, Lean, and Six Sigma were becoming popular. Taguchi was a well-known name and using robust design, design of experiments, and statistical process control (SPC) was discussed regularly in the industrial press. On further examination, however, it became apparent to me that while these tools were recognized as beneficial and important, very few companies were applying them efficiently; furthermore, there was no overarching methodology to guide teams on where and when they should be applied. I spent the next few years developing an understanding of the fundamental needs of product development and manufacturing organizations and the shortcomings of their variation risk management programs. I visited over 30 companies producing a wide range of products. From this emerged formalized programs and new tools to assist in variation risk management processes.

In the spring of 2000, a division of a global defense company approached me. Its division wanted a key characteristics process that could be applied during product development to avoid expensive rework in production. At the same time, Analytics Operations Engineering, a manufacturing operations consulting firm, offered me a position that would allow me to apply the tools I had developed to companies. Faced with these complementary opportunities,

I took a year's leave from MIT to implement my work in industry. One year and several exciting client projects later, I decided to leave academia to work full-time in the application of variation risk management in industry.

This book is the result of my academic research on a variety of industry practices and my experience as a consultant applying them to real-world problems. The tools are geared toward product development and manufacturing organizations. The book is written to address a wide audience from the management level to the practitioner. It is designed to guide readers on how to better focus their quality improvements on what matters and to get the most out of their programs.

I wish to thank all the companies that have participated in my research and have been my clients. Ideas for many of the tools and techniques in this book were sparked by discussions with people with years of experience in the trenches. I am indebted to them for their time and input. I want to thank all companies I have worked with, including Baxter Healthcare Corporation; The Boeing Company; Boston Scientific Corporation; Ford Motor Company; Eastman Kodak Company; Hewlett Packard Company; Honeywell International Inc.; Johnson & Johnson; Lockheed Martin Corporation; NASA; Northrop Grumman Corporation; Xerox Corporation; and the U.S. government through the Fast and Flexible Communication of Engineering Information in the Aerospace Industry Project, the Lean Aerospace Initiative, and the Center for Innovation in Product Development at MIT. In addition, thank you to my previous and current graduate students: Your research and insights were invaluable. The feedback and help I received from Steve Bandler, Joy Campbell, Ian Crookston, Christopher Hull, Mark Hurwitz, Kent Kuiper, Earll Murman, and Stephen Walls were helpful in shaping this document. Finally, thanks to Dr. Roy Thornton the editing, feedback, and support was invaluable. Thanks to Analytics Operations Engineering (www.nltx.com) for supporting the development of this book. Additional papers, references, information and conferences on variation risk management can be found at www.variationriskmanagement.com.

FIGURES

TABLES

TEXT BOXES

NOMENCLATURE

A	Cost of delivering the product at either the upper or lower limit (defined by Taguchi) of the tolerance band
b	Bias (average distance from mean)
C_{ci}	Cost of complaint type i
C_D	Cost of a defect
C_e	Cost of expediting
C_{excess}	Excess cost per unit
C_{exp}	Expected cost of variation (risk)
$C_{(g_p)}$	Cost of gap as produced
C_l	Cost if the gap is less than zero and significant rework is required
C_m	Cost of materials in a subassembly
C_p, C_{pk}	Capability measures
C_{rework}	Cost of rework
C_{si}	Cost of setup at station i
$C(x)$	Cost of variation as a function of the mean
g	Gap
g_p	Gap as produced
h	Height
I	Current
L	Loss as defined by Taguchi, equivalent of C_{exp}
L_i	Length i
LSL, USL	Lower and upper specification limits on a tolerance
n	Number of stations
n_{ci}	Number of complaints of type i over a fixed time period.
n_i	Number of parts in inventory
N_b	Number of parts in a batch
p	Defect rate
p_{ci}	Probability of complaint type i
p_N	Normal cause defect rate
p_s	Special cause defect rate

$p(x)$	Probability density function
P_i	Contribution of KC i
r_i	Rework rate at station i
R	Resistance
s	Slope of the cost curve
S_i	Sensitivity
t_i	Time
V	Voltage
x	Value of a characteristic
y	Yield
y_{bi}	Percent of batches with zero rejects
y_i	Yield at station i
α	Angle
Δ	Tolerance width
Δx	Deviation from the mean
μ	Mean
σ	Standard deviation

ACRONYMS

AQS	Advanced Quality System
CAD	computer aided design
CPU	central processing unit
CR	customer requirement
CSR	critical system requirement
CTQ	critical-to-quality characteristic
DFA	design for assembly
DFM	design for manufacturing
DFSS	Design for Six Sigma
DMAIC	define, measure, analyze, improve and control
DOE	design of experiments
DPM	defects per million
DPMO	defects per million opportunities
FAA	Federal Aviation Administration
FDA	Food and Drug Administration
FEA	Finite Element Analysis
FMEA	failure modes and effects analysis
FTY	first-time yield
GD&T	geometric dimensioning and tolerancing
HVAC	heating, ventilation, and air conditioning
I-A-M	identification, assessment, and mitigation
IPPD	integrated product and process design
IPT	integrated product team
IV	intravenous
JIT	just-in-time
KC	key characteristic
LCD	liquid crystal display
LSL	lower specification limit
MTTF	mean time to failure
MTTR	mean time to repair
NC	numerically controlled

NTSB	National Transportation Safety Board
NVH	noise vibration and harshness
P/E	price to earnings
PCB	printed circuit board
PCD	process capability data
pdf	probability density function
QC	quality control
QFD	quality function deployment
R&R	repeatability and reliability
ROI	return on investment
RSS	root-sum-squared
RTY	rolled throughput yield
SOP	standard operating procedures
SPC	statistical process control
TQM	Total Quality Management
USL	upper specification limit
VRM	variation risk management

1

INTRODUCTION

The increased rate of change of technology and the speed at which companies now can adopt new technology has reduced the "new toy," or first-to-market, effect in many industries. Today, consumers take for granted that every company will incorporate the newest available technology in its products. Customers want the newest features while getting the best value for their money. This is true whether the customer is a purchaser of medical devices for a health maintenance organization, a government agent purchasing multi-million-dollar defense equipment, or an everyday consumer of digital cameras. Because of these customer expectations, *operational efficiency*—the ability to quickly and reliably design and produce high-quality products at low cost—is now a major differentiating factor between companies. Operational efficiency drives profit, return on investment, and ultimately, shareholder value. It is companies that can bring well-designed new products to market quickly, operate efficiently with minimal overhead, and produce high-quality products with minimal scrap or rework that will succeed and grow.

A critical factor in operational efficiency is the ability to design and build high-quality products. This book focuses on one aspect of quality: variation in product dimensions and features and its impact on the performance, cost, and safety of a product. Variation in production impacts many aspects of operational efficiency (Fig. 1-1): inventory, touch time, warranty and product returns, and capacity utilization. Ultimately, variation increases waste and reduces profit.

When observing quality initiatives of many companies, one sees an interesting contradiction emerge. Most companies talk about the importance of

1

Figure 1-1. Lost profit due to variation.

designing and building high-quality products that are robust to variation and produced at a low cost. They can quote fluently such authors as Crosby, Deming, and Taguchi (Crosby, 1979; Deming, 1986; Taguchi, 1992) and many have Six Sigma Black Belts on staff. However, when actually designing a product, they say, "Robust design is great, but we cannot apply it everywhere. We just don't have time." Schedule takes precedence, and consequently, manufacturing often is handed a product that is difficult to build reliably with a high degree of quality. The question arising from this contradiction is: "If companies understand variation reduction tools and the need for them, why do companies continually struggle to apply them?"

A similar contradiction is apparent in regard to products already being manufactured. Organizations understand the need to improve product quality to save cost. They have many tools in place to do variation reduction including Six Sigma, continual process improvements, Total Quality Management (TQM), and statistical process control (SPC), to name a few, but after an initial push do not continue the efforts. They do provide adequate support to help employees understand where to apply the tools. People identify Six Sigma projects without questioning where variation is showing the most impact across the organization. Often the easy projects are identified—not the right ones. In some cases, companies are able to show polished presentations of current quality improvement projects; however, it is often the case that few projects are completed; or, if they are, returns are never as large as promised.

In some cases, implications of a change on the whole organization are not understood and improvements in one area increase costs in others. For example, measurements can be added to control variation; but the step creates a bottleneck in the factory, increasing cycle time and increasing back orders. Although the cost of variation for one group goes down, the total cost to the organization goes up. Other companies with successful Six Sigma programs can demonstrate that their projects have a positive return on investment but can not demonstrate that the projects they picked were the *best* projects to work on. Given the cost of deploying a Six Sigma or variation reduction program, it is a waste to not assign teams to projects with the highest return on investment and ensure the projects are completed.

In a large range of companies—from automotive to aerospace to medical to component manufacturers—familiarity with tools is not the problem. Three interrelated challenges face companies that wish to successfully apply these tools. First, companies often do not know how to apply limited

---

Why Is Product Development Difficult?

Several quality experts have stated that "quality is free" (i.e., any improvement in quality will return a positive return on investment (ROI)) and that it is less costly to build a good product than to build a defective product. This is true in the production environment. Once a product is in full production, it is easy to identify problems, fix them, and quantify the returns. Most improvements will have a net positive return. However, in design the cost/benefit tradeoff is not as clear.

It is much more difficult and costly to design a good product than to design a bad one. Good product development requires thought, many iterations, and the willingness to challenge your immediate assumptions about the best solution. In addition, designing for cost-effective quality involves multidimensional trade-offs. The design team needs to simultaneously consider the product requirements, cost, manufacturing processes, and variation. Teams also need to be supplied with the right tools and data to be able to successfully balance these often competing requirements. Work done to remove the sources and impact of variation in the design phase can have a significant and beneficial impact on the overall quality and cost of the product. However, it is often not possible to quantify the benefit derived from the effort spent during the design process.

resources effectively. Schedules dictate time spent in product development. In production, a limited number of people are allocated to manufacturing process improvement. Second, teams do not know how to agree on the most critical issues and come to a consensus on how to address them. As a result, organizations tend to operate in a fire-fighting mode and have too many ongoing projects that are rarely finished on time. Third, quality is typically addressed on a part-by-part basis rather than optimizing quality and its impact across the entire organization.

To address these challenges, the methodology entitled *variation risk management (VRM)* was developed. VRM refers to the proper allocation of limited resources to variation control efforts in order to improve quality and reduce cost as efficiently and effectively as possible. In other words, given a large number of opportunities to apply variation control and reduction tools, variation risk management identifies the best opportunities on which to focus. The VRM methodology is based on two fundamental concepts: (1) a *holistic view of variation* and (2) the *identification, assessment, and mitigation* (I-A-M) process.[1] Variation risk management can be applied either proactively in the product development process or to an existing product. Variation risk management requires the integration and participation of all functional groups that have influence over product quality, including design engineering, manufacturing, quality, system engineering, customers, procurement, and suppliers.

The existing literature on variation and quality is substantial. A simple search returned 94 books on Six Sigma, 43 books on product development and quality, and 151 books on SPC some of which are listed in the bibliography. Given the amount of material on variation reduction, why write another book? Most books on the market talk about applying specific techniques such as Six Sigma, SPC, robust design, variation reduction, and manufacturing process improvement. Still others promote the importance of quality and reducing variation, but only provide minimal guidelines on the technical and practical aspects of implementing a program.

This book presents proven quantitative methods that have helped many product development and production organizations effectively and efficiently reduce the total impact and cost of variation. The VRM methodology has helped companies identify key areas for improvement, significantly reduce cost, and improve quality. It has been used to guide where Six Sigma, robust design, and other quality improvement efforts are applied to get the best re-

---

[1]The term *process* has many meanings in this book. We will differentiate by describing what type of process we are discussing—for example, manufacturing processes (those that are used to build the product) versus product development processes (those used to design the product).

turns. The tools in this book have been used on a wide range of products and technologies including aircraft, automotive, engine design, medical devices, electromechanical devices, optical assemblies, printed wiring boards, and microelectronics.

This book is fundamentally different from other quality and product development books in that it presents quantitative methods, teaches you to look at a product as a system rather than as individual parts, shows you how to target limited resources, and provides methods for ongoing tracking and reduction of variation costs. It provides a fact-based and rigorous approach enabling teams to agree on priorities. One goal of this book is also to demystify the "magic" used by consultants to analyze organizations. It will teach the reader how to think holistically about variation in products.

Some tools presented in the book are good engineering put in a clear framework. This book will help the entire organization to behave in the same way that your best engineers do. In addition new tools that have not appeared before are presented.

1.1. THE COMPETITIVE ADVANTAGE OF VRM

The Ability to Reduce the Impact of Variation Is Critical to Being Competitive. Companies that minimize variation successfully will see lower costs and higher customer satisfaction. CEOs and Wall Street value bottom line improvements because savings accrued by a company can have a large impact on shareholder value. For example, if a company trades at a P/E (price to earnings) ratio of 10 to 1, a recurring saving of $1 million after tax can increase total equity value of the company by $10 million. An organization can use variation risk management tools to:

- **Develop Better Products.** As has been pointed out in a large number of books and articles, the product development team has the greatest control over both cost and quality early in product development (Fig. 1-2). The earlier that variation is used as a criteria to select concepts, parts, and manufacturing processes, the better. When teams address variation late in the product delivery processes, their choices and options for reducing the impact of variation are more limited, are more expensive, and have a smaller impact. The VRM methodology helps teams efficiently identify the best opportunities to improve the design early, resulting in improved customer satisfaction, less scrap and rework, fewer defects, and a leaner production process.
- **Avoid Excessive Precision in Parts.** To avoid potential defects, it is tempting to include expensive precision parts "just in case." Tightening

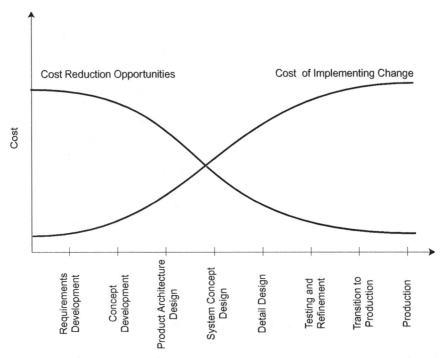

Figure 1-2. Cost of change versus ability to affect change during product development.

the specifications limits is often used to mitigate quality issues, but this approach often adds excess costs and decreases operational efficiency. Teams must have a thorough understanding of the acceptable level of variation in order to choose the appropriate quality level of parts and manufacturing processes. The VRM methodology identifies parts that have the largest impact on product quality and allows the team to effectively prioritize where additional precision is required and identify alternative, less costly solutions.

- **Allow Faster Transition to Production.** Transition to production is the stage in product development when the product is introduced into the factory and production rates are vamped to full production. During this time, the team works out any bugs in the product and manufacturing processes. It is expensive to redesign at this stage because equipment and tooling has already been purchased and installed. Proactive variation risk management can reduce the number of yield and quality issues that surface during transition to production as well as reduce the time to identify and fix unexpected issues.

- **Continually Improve Product Quality and Reduce Cost Throughout the Life Cycle of the Product.** Companies can lose market share because of a real (or perceived) problem with quality. Proactive variation risk management will help organizations improve product quality before it gets into a customer's hands. In addition, VRM helps identify why products are being returned and links customer return data to manufacturing processes and, best of all, to product design.

- **Reduce Rework and Scrap.** Rework and scrap waste time and resources without providing long term benefits. The occurrence of scrap and rework indicates potential problems, which include a less than robust product design or excessive variation in production processes. Not only do rework and scrap increase the average labor content of a product; they also impact flow through the factory. VRM ensures, through good design, that the product will have a lower scrap and rework rate. During production, VRM identifies key opportunities for reducing costs and quickly implements solutions.

- **Increase Interchangeability and Replaceability.** In military and aerospace applications, the term "interchangeable and replaceable" refers to the ability to replace parts or systems on a product without rework tuning or custom fitting. For example, historically, doors on aircraft were built to match individual aircraft. If a door got damaged, it was not possible to simply install a new door; it had to be custom-fit specifically to the plane. Low variation in mating parts, and robust assembly schemes reduce the need for individual part adjustment, thus reducing the labor required in the manufacturing process and decreasing ongoing maintenance costs. Improving interchangeability is especially important for airlines, where delayed or grounded aircraft cause great expense and poor customer satisfaction.

The Ability to Utilize Limited Resources Effectively to Control the Impact of Variation Is a Competitive Advantage. Variation control tools are not new, nor are they proprietary or patentable. Every company has access to the same tools and methods; any company can develop or purchase a Six Sigma program or hire an SPC expert. Using variation reduction tools only brings a company to the same level playing field. If executed poorly, these programs can have no effect at best, and at worst, negative effects such as delaying the launch of a design or consuming significant resources in production.

Companies must ask themselves, "Given that my competitor has access to the same tools, how do I apply them more effectively, get better results, and spend less money?" One way to improve operational efficiency is to use variation risk management for planning and execution. This means making decisions about design concept, manufacturing processes, and part tolerances that minimize the cost, performance, and safety impacts of variation at the lowest cost with minimal impact to the product development schedule. In production, this means (among other things) picking projects that will have the highest return on investment for the entire enterprise (not just one function) and executing them to get returns quickly. In summary, organizations need to manage variation, not just reduce it.

1.2. GUIDE TO READERS

This book teaches the reader how to determine where to apply variation reduction and quality tools given a limited set of resources. It describes quantitative tools and methods (illustrated with examples) that will provide a competitive advantage in developing, producing, and delivering products. These methods can be applied during either product development or production. Ideally, variation risk management should be included in the way of doing business rather than as a separate program. This book provides guidelines on the implementation of corporate-wide variation risk management programs. However, individuals or teams looking for tools and methods to apply immediately can use individual chapters and subsections to guide their efforts. This

---

Low versus High Volume—Does VRM Apply?

While many products are produced at high volume, there is also a wide range of products produced at a low volume—for example, aircraft and satellites. A low production volume is often used as an excuse not to apply variation management tools in design and production. In fact, variation risk management is as or more critical in low-volume products for several reasons.

First, low-volume products are typically more likely to require significant rework and tuning to perform correctly. Rework and tuning are expensive, especially considering the cost of the parts and labor. Second, products that need to be tweaked in the factory are more likely to experience field failures because they typically function on the edge of their operating ranges. Third, often technology and processes developed for one low-volume product will be reused for future product generations. While the initial production volumes are low, the end result is a relatively high volume. Because of these factors, ensuring a design that is insensitive to the existing process variation is critical to the cost, performance, and quality and ultimately the profitability of any low-volume or high-volume product.

book assumes that the organization is using an integrated product team (IPT) approach to product design where representatives from all functional groups are involved throughout the product development process (see Chap. 10).

What should you expect to get out of this book?

- A set of tools for identifying where to apply variation reduction techniques and guidance on what techniques to apply
- The opportunity to use real data from your factory, previous designs, and financial reports to prioritize efforts and create a plan for addressing your specific variation issues
- Project management guidelines that will keep your team on track and focused
- Knowledge of common pitfalls so you can avoid them
- An implementation template from which you can design your own VRM program

What types of people will be helped by this book?

- **Those Who Have Been Told to Implement a Variation Reduction Program (VRM Implementation Leaders).** Often, a small team is tasked with identifying and developing a company-wide program for implementing a variation risk management program. This book provides necessary information for developing such a program.
- **Product Leaders Who Want to Apply VRM to a New Product (Product Leaders).** This book provides a step-by-step guide to integrating variation risk management into a product development process.
- **Plant Managers Interested in Reducing Production Costs and Improving Quality (Production Leaders).** This book provides a step-by-step guide to implementing a variation risk management program for products already in production. It will teach techniques to track ongoing quality costs and identify key areas for improvement.
- **Those Who Have Been Asked to Identify Six Sigma Projects (Six Sigma Champion).** This book helps identify where to apply Six Sigma projects to get the most benefit out of limited resources.
- **Members of Integrated Product Teams Asked to Apply VRM Tools during Product Development (Product Development IPT Members).** This book gives details of the VRM methodology and specific tools and methods needed to execute it.
- **Members of Teams Requested to Reduce Costs on an Existing Product (Production IPT Members).** This book provides basic tools and methods that are used to execute production improvements.

The book can be used at many different levels and for several purposes. Each chapter provides a broad overview and guide as well as very specific tools and methods. Most chapters are divided into two sections: one for those using VRM in product development and one for those using it on products already in production. Reader types listed above will want to read different sections depending on their needs and level of knowledge. Figure 1-3 shows a map pointing different reader types to the most important chapters for them. A filled-in dot (●) indicates that the reader should read the section in detail, a circle (○) the sections the reader could read in less detail. The sections with no marker may contain information of less interest to the given reader. It is highly recommended that readers identify which chapters should be read first and begin with those.

This book is divided into three sections. The first section (Chaps. 2–8) provides an overview of the entire VRM methodology and details the identification, assessment, and mitigation procedures. The second section (Chapters 9–11) describes organizational structures and requirements for implementing a VRM methodology. Appendixes A, B, C, and D discuss information and implementation tools that can help in deploying VRM within an organization. Each chapter begins with an introduction of the basic concepts covered and then describes, through examples, how to apply them to either a new design or an existing product in production.

The subject matter of the chapters is as follows:

- Chapter 2 introduces basic concepts of variation, a holistic view of variation, and the three-step process of identification, assessment, and mitigation. In addition, it introduces the application of variation risk management in product development and production.
- Chapters 3 through 8 cover the three procedures in the VRM methodology (identification, assessment, and mitigation). Tools and methods for each procedure are presented along with examples.
- Chapter 9 describes how VRM should be integrated into product development, transition to production, and production.
- Chapter 10 provides a list of tasks and responsibilities for each member of the organization, including in-house team members as well as suppliers.
- Chapter 11 provides a structured approach to implementing a VRM program in an organization.
- Chapter 12 reviews the book highlights.
- Appendix A provides a benchmarking tool, the VRM maturity model, used to measure the weakness and strengths of an organization's VRM process.
- Appendix B describes how to build a database of process capability information that can be used to select processes and assess the capability of achieving tolerances.

Chapter	Reader type	VRM implementation leaders	Product leaders	Production leaders	Six Sigma champion	Product Development IPT members	Production IPT members
2: Basics of Variation Risk Management		●	●	●	●	●	●
3: Identification	Introduction	●	●	●	●	●	●
	Body		○	○	●	●	●
4: Overview of Assessment		●	●	●	●	●	●
5: Assessment of Defect Rates	Introduction	○	●	●	○	●	○
	Body	○	○	○	○	●	○
6: Assessment of Cost and Risk	Introduction	○	●	●	●	●	●
	Body	○	○	○	●	●	●
7: Assessment of the Quality Control System	Introduction	●	○	●	●	○	●
	Body	○	○		●	○	●
8: Mitigation	Introduction	●	●	●		●	●
	Body	○				●	●
9: Integration of Variation Risk Management with Product Development	Introduction	●	●	○		●	○
	Body	●	●	○		●	○
10: Roles and Responsibilities in Variation Risk Management		●	●	●	●	●	●
11: Planning and Implementing a Variation Risk Management Program	Introduction		●	○			
	Body	●	○				

● Reader should spend time to read in detail.
○ Reader can skim to get main points.

Figure 1-3. Chapter map.

- Appendix C describes how VRM can be integrated with other improvement initiatives such as Six Sigma, Lean, and Kaizen.
- Appendix D provides copies of the flow diagrams for the identification, assessment, and mitigation procedures.

2

BASICS OF VARIATION RISK MANAGEMENT

Traditionally, the quality control (QC)[1] function was primarily responsible for issues relating to variation. QC's tool kit was typically limited to testing and inspection to prevent defective product from leaving the factory. The product development group traditionally handed over a new product design expecting manufacturing to hold to specified tolerances and work out any production bugs caused by design. The manufacturing organization was expected to get new products out on schedule and meet set production rates. Finance's task was to manage costs without being involved in product development, manufacturing process improvements, or design changes. Marketing developed new customers and ascertained needs for new or modified products, sometimes promising features and quality levels that outstripped technology and production capabilities. Changes in the competitive landscape have made this way of doing business untenable.

In the last few decades, many organizations have changed their quality philosophies and have made reducing variation and its impact the responsibility of the entire enterprise. Even with this philosophical change, most quality tools focus on meeting customer requirements (CRs), achieving part specifications and tolerances, or improving specific manufacturing processes. Tools that focus only on CRs do not help the team determine how to achieve targets. Tools that focus on parts and manufacturing processes do not help to identify what specifications, tolerances, and manufacturing

---

[1] Also called quality assurance or quality engineering.

processes are critical to CRs. The difficult part of managing variation effectively is linking these two areas of effort.

Variation is inherent in all processes, natural or man-made. Sometimes variation will result in increased cost, reduced performance, defects, or customer dissatisfaction. The tools to address variation fall into two categories: (1) those that reduce the sources of variation and (2) those that reduce the impacts of variation. Both categories of tools can be applied during either design or production. These two classes of tools are discussed throughout the book.

- **Reducing Sources of Variation.** These approaches work to decrease the amount of variation introduced by manufacturing processes, suppliers, operations, and so on. These tools are broadly termed variation reduction.
- **Reducing Impacts of Variation.** Different design concepts, product architectures, technologies, build sequences, tooling, and layouts can impact how sensitive the product is to sources of variation. These tools are broadly termed robust design.

Section 2.1, basic principles of variation risk management, introduces how VRM is used in both product development and production, and the identification, assessment, and mitigation procedures. Section 2.2 reviews basic concepts of variation. All readers should read Sec. 2.1, but they only need to read Sec. 2.2 if they require a refresher.

2.1. BASIC PRINCIPLES OF VRM

The VRM methodology is divided into three phases: identification, assessment, and mitigation. Successful application of I-A-M leads to selecting the right problems and executing the right solutions to utilize limited resources efficiently. During selection of the right problems, the team must be:

- **Holistic.** The ability to see how customer requirements are delivered by the individual parts and manufacturing processes is critical to selecting both the right area to work on and the right solution.
- **Process Oriented.** In determining the right problem and right solution, the team must go through a reliable and repeatable VRM methodology. Using identification, assessment, and mitigation will increase the likelihood that the team focuses on most productive areas.
- **Data Driven.** Quantitative methods should be used to identify what

areas of the product to work on and what solution to apply. These models should be populated with "real" data from production (i.e., process capability data).

To apply the right solutions, the team must be:

- **Timely.** Teams should apply the VRM methodology as early as possible in the life cycle of a product. By starting execution as early as the requirements definition phase, the best and most cost effective solutions can be applied. By waiting until late in product development, the team limits its available tools and mitigation strategies.
- **Effective.** Teams should use good project management to execute the projects identified using the VRM methodology.
- **Efficient.** The team should use its resources wisely. The right level of analysis should be used at every stage. For example, the team probably should not build detailed Finite Element Analysis (FEA) models during concept design.

The following sections review the three necessary characteristics of variation risk management: holistic, procedure oriented, and data driven. Later chapters provide tools to aid in the timely, effective, and efficient application of variation risk management.

2.1.1. VRM Must Be Holistic

Merriam-Webster's Collegiate Dictionary (tenth edition) defines *holistic* as

> Relating to or concerned with wholes or with complete systems rather than with the analysis of, treatment of, or dissection into parts.

The holistic view of variation looks at the relationships between the parts and how they drive product quality. In addition, the holistic view incorporates the inputs from all functional groups.

When parts are assembled into products, variations in dimensions or features combine to affect performance. It is the combined net impact of variation on customer requirements that results in excess rework (to bring product requirements into specification), scrap, and customer complaints. For example, if variation in wall strength causes a leak in a medical product, the product will be scrapped when it is leak tested. Defects in the wall can be caused by multiple interacting sources including wall thickness, incoming material, manufacturing processing, or handling. Another example occurred with the author's last laptop computer. Variation in solder joints combined with regular abuse of the computer resulted in an intermittent failure of the LCD (liquid crystal display) display. A good shake or whack to the side would usually reset the connector. Ultimately, the toll on the machine was too much, and it refused to work one day. The author now owns a different make of laptop that is much more robust to vibration and abuse. It is irrelevant to the author whether the solder joint falls within the allowable distribution or if connectors are within tolerance; the LCD should not fail.

The holistic view of variation is needed because, while it is defects such as leaks and screen outages that are of interest to the designer and customer, the producer only has control over individual manufacturing processes (soldering and incoming materials). In some cases it is possible to bring a product into specification by means of rework or tuning after it is built; in most cases, however, you get what you build. In order to reliably build products that meet requirements, it is necessary to understand and control the manufacturing processes that drive variability. Specifically, teams need to understand how variation in parts impacts the final product quality (product-to-part interactions) and how individual organizational functions should interact to help reduce the amount and impact of variation (cross-functional interaction).

Product-to-Part Interactions. Complex products are usually composed of many assemblies and parts. Because of this, proposed new products must be broken down into individual parts or assemblies, each of which is designed by different individuals or teams. The breakdown of a complex product into work packages is done as a matter of course in most companies. The difficulty comes in integrating these work packages in production and reliably producing a high-quality product. Variation in the individual parts is not the problem; it is how variations in parts and manufacturing processes combine to impact product performance. Unfortunately, complex interactions are often not identified until the product is put into production, when changes are very expensive.

An example of this is the gap between the door of a car and the car body. Most of the time, these two assemblies are designed by different functional groups, each with its own expertise. It is when the two assemblies are mated that defects arise. The producer and customer only care about the end result (appearance, leakage, or noise). Because defects are caused by the interaction of multiple sources of variation, it is vital to understand how variations in multiple parts, assembly methods, and manufacturing processes relate to each customer and product requirement.

By understanding the how quality is delivered and what manufacturing processes contribute the most, the organization can focus limited resources on the critical few processes. Without the holistic view, selecting the critical processes is guesswork and may result in not identifying the most effective manufacturing processes to control or improve.

Another way to explain the need for not addressing variation on a part-by-part basis is to look at the tolerancing process. Traditionally, designers set part and process tolerances based on previous designs and/or on engineering judgment. *Tolerance allocation,* done correctly, systematically allocates the allowable tolerances and/or engineering margins in the product requirement to the individual parts and manufacturing processes (Fig. 2-1). In other words, tolerances are consistent with the engineering intent of the product.

In order to properly allocate tolerances the product development team must understand the relationships between individual parts, dimensions or characteristics and customer requirements for the final product. In addition, tolerances on parts and manufacturing processes must be consistent with current process capabilities. *Tolerance analysis* includes the process of rolling up the expected process capabilities to check whether the allowable margin or tolerances on the product or requirements can be achieved. Tolerance analysis can be done through extreme value analysis, by root-sum-squared (RSS) analysis, Monte Carlo simulations, or by using specialized software (Chap. 5). The holistic view of variation enables both tolerance allocation and tolerance analysis; without it, tolerancing is done by guesswork and may result in lower quality and/or higher process costs.

Cross-Functional Interactions. The holistic view of variation bridges functional groups within an organization. A single functional group alone cannot determine costs and risks associated with variation, nor can a single

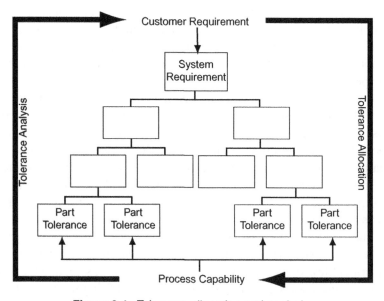

Figure 2-1. Tolerance allocation and analysis.

group fix the product alone. All functional groups (design, manufacturing, quality, suppliers, finance, etc.) must work together during new product development and in continually improving existing production. Typically, this is done in an integrated product team.

Coming to a consensus about the critical areas that require improvement is the first challenge. Coming to the consensus includes developing a common metric for cost of variation that measures the impact of variation across the entire organization. The second challenge is implementing the fixes. In many cases, the functional group that executes the change is not the one that will directly benefit from it. When budgets are assigned along functional lines, getting one team to help another can be difficult. For example, the design group may spend extra time tuning a product to make it robust and may exceed its budget. However, their extra effort during product development will reduce manufacturing costs and increase customer satisfaction. Thus, the net benefit to the entire organization is positive. It is necessary for the functional groups to identify these trade-offs and communicate them to management to allow informed decisions to be made. The total cost and total benefit to the entire organization should be quantified and used as the criteria for identifying where resources should be allocated. This book guides teams on how to work together in identifying the best variation reduction strategy for their product. Chapter 10 describes the specific role for each functional group and how teams should work together.

2.1.2. VRM Must Be Process Oriented

In addition to the holistic view of variation, it is important that teams prioritize where the product is at highest risk from variation using a methodical, proven, and consistent VRM methodology. It is always tempting to immediately start to brainstorm solutions rather than to make sure the team focuses on the problems that will have the largest impact on quality, cost, and customer satisfaction. First, teams must come up with a complete list of potential weaknesses in the product. Second, they must prioritize these weakness using data-driven methods and identify the highest-risk areas. Finally, teams should evaluate various approaches to mitigating variation and determine the best strategy and execute the solution. By going through this step-by-step VRM methodology, teams are more likely to identify the best opportunities, determine the optimal response, and obtain the desired results.

This book will walk the reader through the three-step method of identification, assessment, and mitigation (Fig. 2-2). The I-A-M procedures comprise a methodical prioritization approach to variation risk management. By rapidly identifying critical issues with the highest cost and/or risk, I-A-M ensures that teams focus on product areas that have the largest impact on

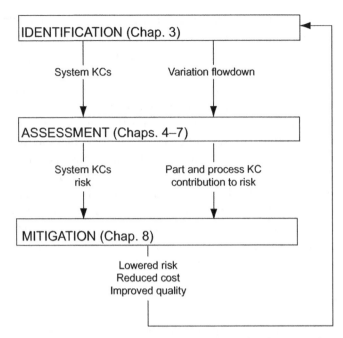

Figure 2-2. Identification, assessment, and mitigation procedures.

cost, performance, and quality. The I-A-M procedures can be applied during either product development or production. Ideally I-A-M should start during product development and be iteratively applied at each stage in product development to ensure that a product is optimally producible (Chapter 9). However, applying the VRM methodology during product development does not let you off the hook in production: Quality and cost should be continually improved throughout the life of the product.

Identification (Chap. 3). Even the simplest product has hundreds or thousands of characteristics, parts, and dimensions that define it. However, only a small subset of these characteristics are linked to customer requirement that are sensitive to variation. For example, the acceleration of a car may be important to a car owner, but small variations introduced during production may not affect acceleration in a noticeable way. However, variation in a significant number of engine parameters will impact a car's noise vibration and harshness (NVH), and a driver will notice any difference between cars. Identification is a methodical procedure for determining which requirements are likely to be impacted by variation and tracing them to the lower-level parts and manufacturing processes that contribute to variation. Critical system requirements that are sensitive to variation and their contributors are termed the *key characteristics* (KCs) of the product. KCs can be associated with the product, systems, assemblies, parts, and manufacturing processes. For example, variation in an aircraft's structures and panel shapes can cause skins to meet unevenly. The seam gaps and steps and part dimensions that impact the seam are KCs. The output of the identification procedure is a *variation flowdown* (also called a *KC flowdown*). The team uses this variation flowdown as a framework for assessment and mitigation. For a product in production, the KCs are fairly static. For a new product, the flowdown is updated and altered as the concept and details are specified and finalized.

Assessment (Chaps. 4–7). Once the variation flowdown is created, the next step is to assess the relative risk and/or cost of each KC. Along with the variation flowdown, inputs to assessment include data about variation costs, data on capabilities of manufacturing processes and a model of how variation in the parts and processes impacts the customer requirements. Assessment is used to identify the highest-risk/cost areas of the product due to variation and to identify major contributors to risk. Various assessment tools can be used depending on project maturity, data availability, and whether the project is in the product development or production stage. During product development, availability of good process capability data is critical to assess the performance of any new product (Appendix B). In the product development stage, the goal is to predict the cost and likelihood of producing a defective product. In the production stage, the goal is to measure the total cost of variation. The output of assessment is the relative risk of not achieving each system KC and identification of what parts, manufacturing processes, and suppliers are the major contributors to high-risk KCs.

Mitigation (Chap. 8). Mitigation is the process of reducing either the magnitude or the impact of variation. Once the IPT has identified and assessed a product's key characteristics, it can apply various mitigation strategies such as design changes, manufacturing process changes, manufacturing process improvements, manufacturing process monitoring, or inspection. The first part of mitigation is determining the most appropriate strategy. During product development, the team must identify where focusing on variation and including variation as a critical requirement are necessary. For products in production, the team must select and manage a portfolio of variation and quality improvement projects that most efficiently apply limited resources. The output of mitigation is lowered risk, reduced cost, and improved quality.

While this three-step process seems obvious, teams often fail because they do not follow the steps sequentially. Once a team has identified the product characteristics that are critical and difficult to fix, it is often tempting to begin immediately brainstorming solutions. If the team is lucky, it might identify the most effective path to fixing the problem. More likely, time is consumed that would be better spent on higher-risk KCs or more appropriate mitigation strategies.

Variation Risk Management during Product Development

In order to effectively manage variation during product development, several tools must be used in conjunction with the I-A-M process.

- **Concurrent Engineering.** Before the advent of concurrent engineering, the product was designed first and then manufacturing processes were developed to deliver the product as specified by the drawings. With the pressures of higher quality, lower cost, and faster product development times, it is necessary to develop the design and the manufacturing processes concurrently. Concurrent engineering is especially important for VRM because without an understanding of how the product is going to be manufactured, it is difficult to identify where the product may suffer from variation issues.

- **Integrated Product Teams.** These teams are critical to the successful implementation of VRM. Integrated product teams include representatives from multiple functional organizations who each provide input on the product's design and manufacturing methods. The goal of IPTs is to ensure that the product is designed to be optimal for the organization, not just for one functional group.

- **Stage-Gate Product Development Processes.** A company should have a formal product development processes which is constructed of a stages separated by formalized gate reviews. The I-A-M procedures should be iteratively applied at each stage. Chapter 9 describes a typical stage gate process and how VRM can be integrated in.

When applied during product development, the I-A-M procedures are integrated into the overall product development process. Variation risk management is not a "quality thing"; rather, it is a principle of good design. Its deliverables feed into the product development process, and its steps are integrated with cost studies, marketing studies, engineering analyses, and other product development tasks. I-A-M provides a methodical prioritization process that can be used to identify where the team needs to trade off costs, customer requirements, and tolerance targets for the product and its proposed manufacturing processes. During each stage gate review process, the IPT is evaluated not only on whether it meets the target requirements of the product but also on whether the requirements can be achieved reliably and at a reasonable cost.

Despite the benefits of proactively reducing the impact of variation, the IPT will likely create excuses for not using the VRM methodology, such as:

"We do not have time to do this."

"We need to get our design out."

"What good is improving the product if we miss our launch deadline?"

"My management will not reward me for this."

VRM does not require an IPT to analyze every part of a design to the nth degree. Instead, it directs teams to methodically analyze—throughout product development—the risk associated with production variation, and to find ways to reduce it or its impact. The IPT's choice of detailed analysis or cursory estimates of risk depends on two factors:

1. **Product Development Schedule:** the current state of product development
2. **Breadth of Analysis:** how much of the product is to be analyzed

---

Fire Fighting versus Focused Problem Solving

I learned the importance of focus while throwing french fries to seagulls when I was a child. If I threw one french fry in the air, the seagull could catch it easily. If I threw two in the air, the seagull would miss both as it tried to catch both. If I threw a handful in the air, the seagull would nearly crash into the pier.

The default behavior for most organizations is a fire-fighting mode and chasing too many "french fries." Unexpected problems arise that demand the attention of the entire group. Emergency meetings are held to plan a solution that becomes the subject of significant attention until another emergency arises. The emergencies can either be real or imagined. Often an offhand remark by senior management can be amplified into a critical issue as it is passed down the chain of command.

The problem with fire fighting is that it constantly changes the priorities and focus of the group, preventing the efficient use of people's time. Unfortunately, people are rewarded for putting out fires rather than preventing them.

In addition, teams often have too many projects ongoing simultaneously. The benefit of having too many projects is that, when asked if a problem is being addressed, team members can always answer "Yes." Increasing the number of projects is easy, and every time a fire is started, the entire team jumps on the problem to solve it. The downside is that there is typically a new fire every week, so projects begun are never finished or their completion times are pushed out. Finally, people are often tempted to work on easy projects; however, it is usually the most challenging projects that yield the highest return.

Teams need to learn to chase after one french fry at a time, catch it and then move to the next. Using this approach, projects get completed and the benefits of the investment are accrued to the company.

Early in product development, the analysis scope will be broad, while the detail will be shallow. Analysis early in product development is typically qualitative. The early analysis will identify those areas that require more detailed analysis later. As the product becomes more mature, analysis detail will increase and the team will use more mature tools while breadth will decrease (Fig. 2-3). For example, detailed analysis such as assembly modeling will only be applied to those KCs where risk is high or the amount of risk is uncertain. The team should maintain a constant effort (breadth × detail).

During product development, the teams face many challenges. There will always be a battle between the fixed, one-time cost of effort in product development and the recurring costs in production. Unfortunately, the short-term pain and pressure to just get the design done often wins against ensuring the long term quality and producibility of the product.

In addition, with concurrent activities, there is the chicken or the egg problem, which must be solved through iteration. In concurrent engineering, the manufacturing group has a difficult time commenting on incomplete designs, but the design group cannot design for unknown manufacturing processes. The same challenge faces variation risk management. In order to assess the ability to achieve target tolerances, a design concept and manufacturing processes must be chosen. However, the team should choose the design concept and manufacturing processes to best achieve tolerances. Figure 2-4 shows the iterative concurrent engineering and variation risk management required to obtain the best match between product design and selected production methods.

If carried out correctly, the VRM methodology should reduce the amount of work an IPT needs to do. Ideally, VRM will identify a subset of areas that require significant attention from the IPT, and the team can focus its robust

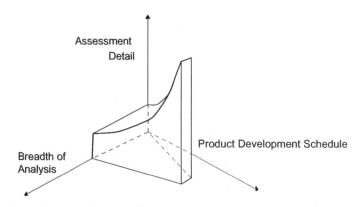

Figure 2-3. Assessment breadth and detail in product development.

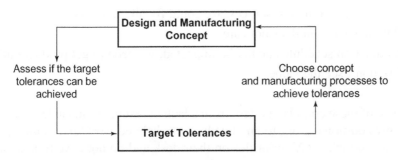

Figure 2-4. Tolerance iteration in product development.

design or design for Six Sigma (DFSS) efforts on this small subset. Because priorities are agreed on, teams are less likely to spend a lot of time and work on changing priorities.

Variation Risk Management during Production

Variation risk management for products in production shares the basic I-A-M framework with VRM applied to new product development; however, organizational issues and VRM tools are different. It is often easier for companies to justify variation reduction during production because return on investment (ROI) can be calculated. Six Sigma has demonstrated benefits that can be accrued by improving existing processes. In addition, there is rarely a schedule constraint that prevents action from being taken. However, the budget, resources, and talent for product and process improvements are often limited.

The VRM tools used in production are based on concrete cost and yield numbers. Rather than predicting the reject or yield rate (as is done in product development), it is possible to accurately assess the current scrap and rework rates and determine the total cost of variation. However, options for mitigation are limited. It can be expensive to make major process or design changes, and teams are often limited to minor process improvements and/or special quality control.

Given the resources, people usually are willing to improve existing production lines. However, they are less willing or able to spend time to analyze the production environment methodically to determine where to best apply variation reduction tools. Typical justifications for resistance are:

"We are already doing that."

"We had a meeting and decided what to do."

"We know what the biggest problems are."

"If we just had more resources, we could fix all critical issues."

"My management does not care."

"Production schedule can not be impacted, we need to get product out."

"We don't see the value."

Most of these complaints stem from a lack of resources and a desire to not add any additional work to an already overwhelmed workforce. Often people have too many "A" priorities on their desk and do not have time to step back and think through what is really critical. Another major issue is that management does not see the value of systematic approaches to variation reduction because of tradition or a lack of knowledge. In addition, fixing high-cost problems may be an admission of a failure to fix them earlier. Often it is easier to deny there is a problem than to address it.

Variation risk management applied during production identifies key improvement activities—those that will provide the biggest return for resources required—and executes those projects efficiently and quickly. After measures of cost of variation are created, the first step in VRM should be terminating many noncritical projects. An organization required to systematically identify, assess, and mitigate quality issues is forced to prioritize its projects and is unlikely to initiate or continue with noncritical projects.

2.1.3. VRM Must Be Data Driven

Throughout identification, assessment, and mitigation, teams need to use data-driven methods. During identification, teams should use information from a wide variety of sources including engineering, quality systems, manufacturing, and finance to help identify all of the potential sources and impacts of variation. For new products in development, information from ongoing analyses should be used along with information from previous programs (quality reports, warranty data, etc.). For products already in production, all sources of cost and waste should be identified.

Assessment is the most data-driven of the three phases. For products under development the team should use quantitative models of performance and populate these models with representative process capability and cost data. These performance models (discussed in Chaps. 5 and 6) enable a team to predict defect rates and quantify the cost of variation. The phrase *garbage in, garbage out* should be thought of when developing models, populating them, and acting on the results. It is very easy to tweak models to make them return the answer you want. The closer data and models are to reality, the more accurate the answers and the more likely you are to act on the right issues. For products in production, all sources of cost, scrap, rework, labor, capacity losses, and inventory costs should be combined to quantify the total cost of variation.

The selection of mitigation methods should also be data driven and based on quantitative measures of return on investment as well as risk, schedule, and resource availability.

2.2. VARIATION AND ITS IMPACT ON QUALITY

This section introduces some of basic concepts of variation, engineering tolerances, process capability, and process yield. Readers familiar with this material can skip it.

Variation sources can be broken into three broad categories: unit-to-unit factors, external noise factors, and deterioration.

- **Unit-to-Unit Variation** occurs in every part or product. It is impossible to build two parts or products identically. Even very precise processes introduce some level of variation into manufactured parts and assemblies. If not properly managed, variations in individual dimensions and features combine to make a product that is unacceptable to the customer. Variation has a number of sources including variations in incoming materials, manufacturing processes, and operators.
- **External Noise Factors** are those sources of variation that are outside the control of the manufacturing process. They can include variations in environmental conditions and differences in the way customers use or abuse a product. External noise factors cannot be designed out, but their effects can be minimized through good design. Designers often place warnings on the product or in the user's manual that state "Do not drop" or "Do not subject to high temperatures." However, this does not get designers off the hook. If the user is likely to subject the product to nonoptimal conditions and will not be satisfied if the product fails or if the competitor's product can survive, then the team must ensure the design is insensitive to these external noise factors. For example, after an increased failure rate in some aircraft intercom phones, it was discovered that the flight attendants were using the phone to break up ice in the galley. As soon as a small hammer for breaking up ice was added to the galley, the phones stopped breaking. The only method for addressing external noise factors is to design the product to be less sensitive to them or design out their effect. For example, if the product is likely to be dropped, it should be made rigid enough to withstand a fall or should be filled with rigid foam after assembly to cushion parts against impact. If the product is likely to be

subjected to extreme temperatures, automatic fans can be included or parts and components can be selected that are less sensitive to temperature variations.

- **Deterioration** introduces variation into the product as the product is used. Deterioration is a major cause for warranty issues in products: Parts wear, break, or fatigue and fail. For example, engine parts can wear, increasing noise and vibration in a car. Small defects in materials, excessive clearances, or imbalanced rotational parts can increase the rate at which deterioration occurs. Using more rigid parts, purchasing higher-quality parts, using different materials, or employing more precise manufacturing processes can mitigate deterioration. In addition, deterioration of manufacturing machinery leads to a gradual drift of process capability and thus to greater product variability.

All critical characteristics of a product should have a specification and tolerance. A characteristic can be anything from a dimension to a voltage to the Young's modulus of a material. Specifications communicate the ideal target value to operators, inspectors, and suppliers. In addition to a target value for a characteristic, a specification typically includes upper and lower limits for that characteristic (i.e., the tolerance). Any product or part characteristic out-

---

Defects

Deviating outside the allowable tolerance can have a range of implications. Some may cause annoyances (a leaking window); others can have significant health and safety impacts (the inappropriate inflation of an airbag). There are many terms used to describe systems with excess variation, such as *nonconformances, defects, out-of-specifications,* and so on. Different terms may have different implications depending on the industry.

For consistency, throughout this book we will use the term *defect* to indicate a product that varies outside the acceptable limits. *Merriam-Webster's Collegiate Dictionary* (tenth edition) defines a defect as "an imperfection that impairs worth or utility." Crosby in *Quality is Free* defines a defect as ". . a failure to conform to requirements." The relative costs and impacts of the defects will be determined in the assessment phase.

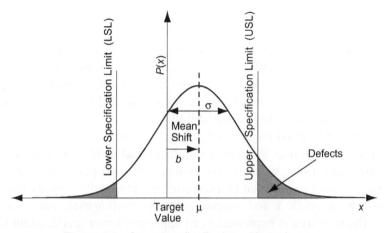

Figure 2-5. Gaussian distribution with a tolerance.

side the upper and lower limits are considered defective. The tolerance is a guide to the manufacturer to set process parameters and quality control requirements. The allowable variation in the customer requirements is allocated to parts and assemblies such that, if the parts' and assemblies' features and dimensions are within tolerance, the customer requirement will be within tolerance.

A manufacturing process will deliver characteristics that will vary from the target specification. The variation of a statistical sample is expressed quantitatively with two values, the *mean value* μ of a characteristic and the *standard deviation* σ of the sample about the sample mean.

Typically, dimensions or characteristics of a statistical sample are modeled according to Gaussian distributions, commonly called bell curves (Fig. 2-5). In a Gaussian distribution there is a high probability of a value being near the mean and a low probability of a value deviating significantly from the mean.[2] For the mathematically inclined reader, the Gaussian distribution is defined by the probability density function (pdf) $P(x)$:

$$P(x) = \frac{1}{\sigma\sqrt{2\pi}}\ e^{-(x-\mu)^2/(2\sigma^2)} \qquad (2\text{-}1)$$

---

[2]Characteristics can vary according to a wide range of distributions (binomial, Poisson, etc.); however, the Gaussian distribution is typically used to model variation and provides a good approximation of the nature of variation. The Gaussian is relatively easy to understand and easy to use in assessment.

where $P(x)\,dx$ represents the probability that a value falls in the range $[x, x + dx]$.

Process capability, expressed variously as yield, defect rate, C_{pk}, or defects per million parts (DPM), is a measure of how well a manufacturing process delivers parts that fall within the tolerance. The percentage of parts that fall outside set tolerance limits is never zero and is a function of the specification limits and process variability.

An easily understood measure of process capability is the defect rate expressed as a fraction or percentage (number bad/number produced). The yield rate, conversely, measures the number of good parts produced. The yield rate, y, can be estimated by predicting the probability density function $P(x)$ of a characteristic x, and comparing the distribution to the allowable tolerance. The tolerance is expressed as an upper and lower specification limit, [LSL, USL].

$$y = \int_{\text{LSL}}^{\text{USL}} P(x)\,dx \tag{2-2}$$

The defect rate p is the fraction of parts that fall outside the acceptable tolerance range and is expressed by:

$$p = 100\% - y \tag{2-3}$$

There are a wide variety of terms to describe various yield calculations. Two typical metrics are rolled throughput yield (RTY) and first time yield (FTY). Rolled throughput yield is the fraction of products that pass through each station without rework or scrap. First time yield measures the total yield including rework. Because a certain number of products are reworked at each station and then passed onto the next station, the RTY will be lower than the FTY. When quoting yields, the team should be very clear on what the number represents or its interpretation may be skewed.

The rolled throughput yield is equal to the product of all station yields y_i:

$$\text{RTY} = \Pi y_i = y_1 \cdot y_2 \cdot \cdots \cdot y_n \tag{2-4}$$

In the case of Fig. 2-6, overall rolled throughput yield is equal to 77.8%. The first time yield is equal to:

$$\text{FTY} = \Pi(y_i + r_i) = (y_1 + r_1) \cdot (y_2 + r_2) \cdot \cdots \cdot (y_n + r_n) \tag{2-5}$$

where r_i is the percentage of parts that are reworked at each station. With rework, first time yield is 91.7%.

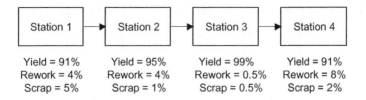

Rolled throughput yield = 91% * 95% * 99%*91% = 77.8%
First time yield= 95% * 99% * 99.5%*98% = 91.7%

Figure 2-6. Production line with yields demonstrating difference between rolled throughput yield and first time yield.

The second method of expressing process capability is to calculate defects per million units. This estimates the total number of defects in 1 million products. Sometimes DPM is used for individual parts or manufacturing processes and sometimes for final products.[3] When looking at yield rates, the difference between a three-sigma capability (99.73 percent) and a four-sigma capability (99.993 percent) may not seem that large. However, the first one will have a DPM of 2700 and the second of 63, a reduction in the number of defects by 97 percent.

The third measure of process capability is C_{pk}:

$$C_{pk} = \frac{\min(\mu - \text{LSL}, \text{USL} - \mu)}{3\sigma} \qquad (2\text{-}6)$$

Typically a KC is considered low risk and acceptable if C_{pk} is above 1.33 and high risk if C_{pk} is below 1.0.

Tables 2-1 and 2-2 show defect rates for different combinations of standard deviations and mean shifts. The first table shows the defect rate and C_{pk} assuming that the manufacturing process on target with different ratios between the tolerance width and distribution. The second table shows the same metrics for a process with a positive mean shift of one sigma.

Process capability can be expressed for short-term, long-term, or special causes of variation (Fig. 2-7). Short-term capability data demonstrates the best-case capability and is typically taken during a single shift when the

---

[3]Defects per million opportunities (DPMO) is another measure used to normalize for product complexity. DPMO computes the number of defects given 1 million opportunities. The more complex the product, the more opportunities for defects. This measure can be biased based on the definition of an opportunity and should be used with caution.

Table 2-1. Defect rate and C_{pk} for centered distributions (bias = 0)

(USL − LSL)/σ	4	6	8	10
C_{pk}	0.67	1.00	1.33	1.67
Defect rate	4.55%	0.27%	0.01%	0.00%
Defects per million	45,500	2,700	63	0.6

manufacturing process is in control. While short-term capability provides a snapshot, long-term capability quantifies process variability over a longer period of time—up to a year. Process capability may vary between shifts, after a shutdown and restart, and over time because of tool wear and other factors. The long-term capability averages together many short runs and the long-term capability is often significantly lower than the short-term capability. The Six Sigma method suggests adding a 1.5σ mean shift to short-term capability to estimate long-term capability; however, there appears to be very little data to back up this assumption.

Special-cause variation induces a significant and sudden degradation in process capability. The degradation can come in the form of either a mean shift or an increased variation about the mean. For example, a broken tool or a new operator may cause a sudden change in the capability of a process.

Table 2-2. Defect rate and C_{pk} for distributions with a positive mean shift (bias = 1.0σ)

(USL − LSL)/σ	4	6	8	10
C_{pk}	0.33	0.67	1.00	1.33
Defect rate	16.00%	2.28%	0.14%	0.00%
Defects per million	160,005	22,782	1,350	32

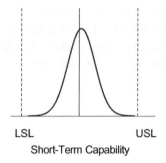

LSL   USL

Short-Term Capability

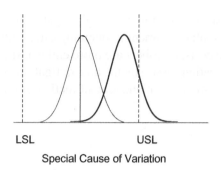

LSL   USL

Special Cause of Variation

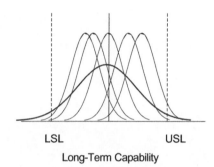

LSL   USL

Long-Term Capability

Figure 2-7. Capability types.

2.3. SUMMARY

Variation risk management requires both a holistic view of product quality and variation and the application of methodical identification, assessment, and mitigation procedures. The identification phase creates a short list of possible variation issues. The assessment phase measures relative risks and costs associated with these issues. The mitigation phase selects and then executes a limited number of projects to improve the product. All three phases are based on quantitative and engineering-based analysis. The I-A-M procedures serve to ensure that the team (1) identifies correct problems to work on and their optimal solutions, and (2) agrees on approaches to take. In addition, I-A-M provides the holistic view of quality that allows for appropriate trade-offs between design and manufacturing costs, between short- and long-term expenditures, and between recurring and nonrecurring costs. This holistic view will allow teams to make choices that are best for the organization, not just for an individual functional group.

Variation risk management should be executed throughout product development and after the product goes into production. While there is often resistance to going through a systematic process, once teams understand that a small investment in planning often reduces the amount of work and life cycle costs, members are more enthusiastic. The next chapters explain the identification, assessment, and mitigation procedures that form the core of the variation risk management methodology.

3

IDENTIFICATION

The first goal of variation risk management is to quickly identify areas where variation has the largest impact on an organization. Variation risk management uses three phases—identification, assessment, and mitigation—to select the critical areas and identify and execute solutions. The identification process, described in this chapter, initiates variation risk management by delivering the holistic view of variation. The identification process enables the team to determine, and document how product quality is delivered and what parts, manufacturing processes, parameters, suppliers, and external noise factors are likely to impact final product quality.

The *variation flowdown* is a system map of variation and the *key characteristics* of the product which the entire team can agree to and provides the framework for assessment and mitigation. The tools and methods for identification are the same for new products in product development and products already in production. The difference between the two is in the information sources used as inputs to identification. All readers should look at the first section of this chapter to understand what key characteristics and variation flowdowns are. The rest of the chapter provides detailed descriptions of how to identify key characteristics and create a variation flowdown.

Figure 3-1 shows the summary of the identification phase and the steps in it. Typically, the identification procedure is done as a team effort. Each team member brings the information outlined in this chapter to a facilitated session where the group jointly executes the identification phase.

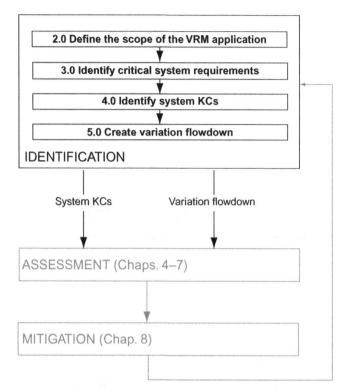

Figure 3-1. Summary of identification phase.

3.1. DEFINITION OF KEY CHARACTERISTICS AND VARIATION FLOWDOWN

To build a complex product, thousands of dimensions, parameters, and characteristics must be specified. These can include diameters, material properties, voltages, and so on. Of these, only a subset are linked directly to customer requirements. For example, only some of the dimensions on spars and ribs of an aircraft will ultimately affect the lift and aerodynamics of the plane. Only a few characteristics from this subset have the potential to impact customer requirements if they vary from their specified values. For example, on an aircraft variation in the spars and ribs can cause steps and gaps between the wing panels that will increase drag and ultimately affect fuel efficiency. The smallest subset is termed the product's key characteristics. The goal of the identification phase is to identify the KCs of the product. Figure 3-2 shows the hierarchy of product dimensions and features.

A key characteristic is different from a critical-to-quality characteristic (CTQ) as used by design for Six Sigma, TQM, and robust design. CTQs en-

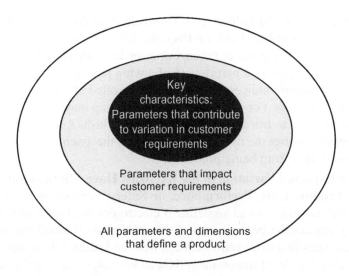

Figure 3-2. Hierarchy of product dimensions and parameters.

compass a larger number of critical issues, some of which are not sensitive to variation and can include specification that will be difficult to design.

The identification procedure should be a team effort. Chapter 10 describes several organizational structures that can be used to identify KCs. The team as a group should identify KCs through first identifying system KCs and then creating a variation flowdown. The rest of this section introduces the concepts of key characteristics and variation flowdowns.

3.1.1. Key Characteristics

The definition of a key characteristic used in this book is as follows:

> A key characteristic is a quantifiable feature of a product or its assemblies, parts, or processes whose expected variation from target has an unacceptable impact on the cost, performance, or safety of the product.

The definition embodies four important concepts.

1. **A KC Target Value and Its Acceptable Variation Should Be Quantifiable.** The assessment phase determines the ability of the existing manufacturing processes and supply base to deliver KCs. The VRM methodology requires yields and costs to be quantified. Thus, identified KCs must have quantifiable target values and quantifiable tolerances to enable assessment.

2. **A KC May Be Identified (and Controlled) at Any Level—Product, System, Assembly, Part, or Process.** KCs can be variation-sensitive parameters of a product (excess drag or lift), an assembly (steps between wing panel), a part (length of a wing panel), or a manufacturing process (fixture indices). KCs can be controlled at any stage of the assembly process. For example, an excessive gap may be filled or shims may be used to bring two pieces of a wing flush. Another option is to carefully control the two parts that make up the assembly to prevent excessive gaps from being produced.

3. **The Expected Variation in the KC Must Have a Significant Impact on Product Cost, Performance, or Safety.** For example, a KC may impact rework costs in a satellite if electronics need to be tuned. A KC may impact the performance of an automobile if small variations in clearances in a drive train increase wear and ultimately result in a decreased life. Most important, a KC can have an impact on the safety of a product. Among other issues, insufficient lubrication of the acme nut screw (NTSB, 2002) was found to be a contributor to the Alaska Airlines Flight 261 crash in January 2000.

4. **The Expected Variation in the KC Must be Likely to Occur.** A simple example is a customer requirement that a pen be blue. It is highly unlikely that the manufacturing plant will put red ink in the pen: As a result, ink color is not specified as a key characteristic. Another example is the seams on a frame backpack. Variation in materials, sewing quality, distance to the edge, and use of the pack (tendency to overstuff it) can impact the time elapsed before the seams begin to fray. The seam strength and all manufacturing processes and dimensions are KCs of the product. (Given the history the author has had with backpacks, variations in parts and manufacturing processes are likely to impact the life of the seams and should be called KCs.)

Other definitions, such as those in the AS9103 standard (SAE, 2001), focus specifically on the features of parts or processes that can be monitored in production. The definition used in this book is broader and encompasses the entire set of features that are impacted by variation.

Another way that is commonly used to illustrate how to select KCs is the Taguchi loss function (Taguchi, 1992) (Fig. 3-3). The Taguchi loss function is a graphical way of demonstrating the impact of variation and importance of keeping manufacturing processes on target. The x axis represents the amount by which a feature deviates from the target, while the y axis represents the cost associated with the deviation. The figure shows that there is an exponentially increasing cost (x) associated with deviating from the target, even for those products that are within the allowable tolerance. For example,

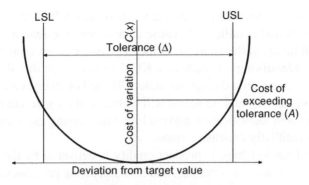

Figure 3-3. Taguchi loss function.

a consumer is unlikely to buy another backpack from the same maker if the seams begin to fail just after the warranty runs out.

While there is still debate about the relationship between the Taguchi loss function and the definition of a KC, the following criteria will be used in this book. It is the combination of the cost of variation and process capability that defines a KC. The product of these two measures is the *expected cost* (Fig. 3-4). Where there is significant variation and it is likely to cause excess cost

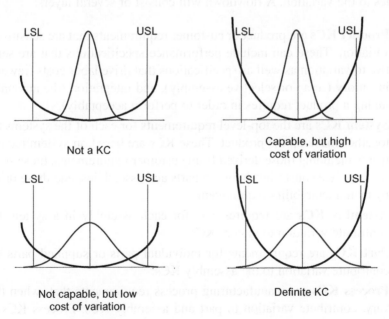

Figure 3-4. KC defined by Taguchi loss function and manufacturing process variation.

the feature can be deemed a KC. Where variation is small in comparison with tolerance limits and is unlikely to cause excess cost, the feature may not be a KC. It will be up to the team to decide. The other cases should be identified as KCs. Identifying a feature as a KC does not mean that the team will be required to take action during mitigation. These borderline cases may rank low on the priority list and no action will be taken. It may be safer to include borderline KCs in identification and exclude them during assessment than to overlook a potentially expensive issue.

As pointed out in Chap. 1, product quality is delivered by the entire system. Variations in parts, assemblies, and manufacturing processes combine to impact requirements important to a customer. The customer cares about wind noise, leakage, and appearance in his or her car, but for a production team those issues translate into complex interactions among the car frame, door, fixtures, materials, and production processes.

3.1.2. Variation Flowdown

The goal in the identification phase is to create a holistic view of how quality is delivered. This holistic view is embodied in the variation flowdown or key characteristic flowdown and captures and diagrams the IPTs knowledge about which customer requirements are sensitive to variation and what contributes to the variation. A flowdown will consist of several layers:

- **Product KCs** are product or customer requirements that are sensitive to variation. These can include performance specifications that are sensitive to variation as well as specifications that drive labor costs (rework), inventory (scrap or selective assembly), and integration (the amount of tuning a product requires in order to perform acceptably).
- **System KCs** are the top-level requirements for each of the systems that together make up the product. These KCs are based on system requirements including those derived from customer requirements, those set by the interface constraints between parts and assemblies, and those driven by manufacturability requirements.
- **Assembly KCs** are requirements for each assembly in a system that contribute variation to system KCs.
- **Part KCs** are requirements for individual parts or supplied parts that contribute variation to the assembly KCs.
- **Process KCs** are manufacturing process requirements that, when they vary, contribute variation to part and assembly KCs. Process KCs include variations in tooling, operators, and machines.

- **External Noise Factors** are sources of variation outside the IPT's control (for example, usage, wear, and environmental factors).

Typically, an IPT will start with system KCs for the system they are designing and create the flowdown. It is too time consuming and complex to create a variation flowdown for an entire product (except for simple products). It will be the responsibility of the system's engineering function to identify product KCs and how they are delivered by individual systems.

The terms *product, system, assembly, part,* and *manufacturing process* as applied to KCs are used broadly to describe the top, middle, and bottom of a variation flowdown. Later, the book will discuss relationships between KCs in the flowdown and will use the following terminology:

- **Parent KC.** This is the next-higher KC in the flowdown. For example, a car's door-to-body seam evenness is a parent KC to the body aperture's shape.
- **Contributing or Child KCs.** These are features that impact variation in a parent KC. For example, the outer perimeter of the car's door, body aperture, and door-body alignment are all contributing or child KCs to seam evenness.

Figure 3-5 shows a generic variation flowdown. The flowdown does not always go from product to system to assembly through to manufacturing processes. Often, the flowdown will skip directly to process KCs depending on the complexity of the system and how it is produced.

Figure 3-6 shows an example of a variation flowdown for a car door. In the figure, the customer requirement is his or her perception of the car door's proper operation. The system being analyzed is the door and door frame assembly. Several characteristics of the door–door frame assembly (system KCs) influence the customer's perception of quality: gaps between panels and door closing force. Each system KC has several contributing assembly KCs associated with either the door or body (e.g., perimeter of the door, and body aperture). Individual parts and manufacturing processes included in the assembly contribute to assembly KCs. The part KCs include the hinge location on the door. Ultimately, system KCs can be traced to manufacturing processes that create features and to external noise factors. In the case of automotive body assembly, fixtures have a large impact on the assembly's variability. External noise factors are not controllable by the factory and include wear, aging, and degradation. For example, seals may age with time and exposure to the elements, resulting in a change in stiffness.

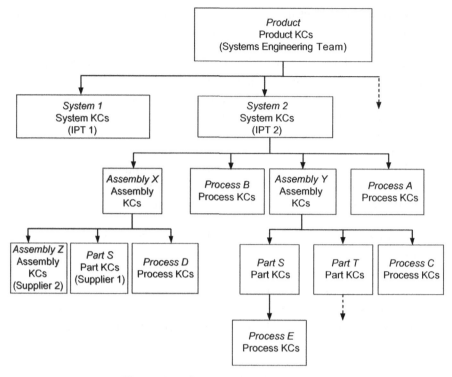

Figure 3-5. Generic variation flowdown.

A team must be able to trace between part and process KCs up to product KCs to identify the following:

- Features that should be controlled
- Where changes in manufacturing processes may impact product quality
- The root causes of quality problems

The variation flowdown process starts with identifying variation-sensitive critical system requirements (CSRs), from which system KCs are derived. CSRs can include requirements set by product functionality as well as those imposed by other portions of the design (e.g., interfaces). The team creates the flowdown by identifying assembly, part, and process KCs that contribute to system KCs. Each branch of the flowdown is stopped once a manufacturing process or a supplied part is encountered. Including suppliers in the flowdown process is discussed in Chap. 10.

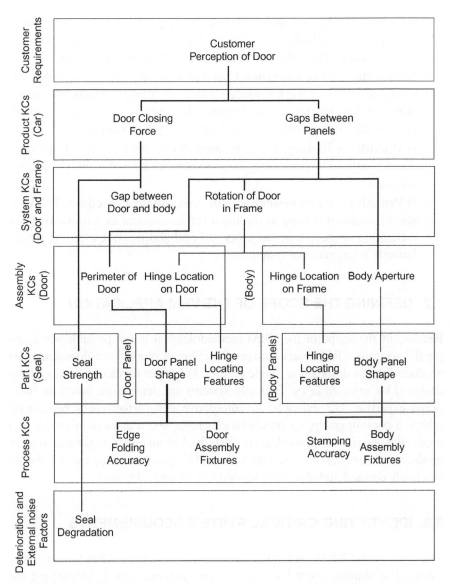

Figure 3-6. Variation flowdown for a car door (Thornton, 1999). Reprinted with permission of Springer-Verlag GmbH & Co.

A variation flowdown is critical to the entire VRM methodology for the following reasons:

- **It Links Customer Requirements (What You Need to Deliver) with Manufacturing Processes (What You Can Control) and Suppliers (What Your Suppliers Should Control).** Understanding this linkage is critical to identifying those manufacturing processes and vendor supplied parts that will have the largest impact on what the product development team and production organization are trying to deliver.

- **It Provides a Map Everyone Can Agree To.** A variation flowdown is a way of capturing the knowledge and input of many functions. Generating the flowdown as a team requires all members to reach a consensus on how the system functions and how quality is delivered.
- **It Provides a History.** It can be used during the transition to production to help with root cause analysis and can be a starting point for a redesign.
- **It Provides the Framework for the Assessment Procedure.** The variation flowdown is used as the input for assessment as a framework for estimating or measuring the costs and probabilities of exceeding the allowable tolerances on system KCs.

3.2. DEFINING THE SCOPE OF THE VRM APPLICATION

Because of the scope of the VRM methodology, it is not possible for a single IPT to apply VRM to an entire product. During product development, most products are not designed as single entities, but rather divided up into systems designed by separate IPTs. Typically systems are determined based on functional expertise. For example, in automotive design there will be, among others, a cooling group, an engine block group, and a drive train group. In a production environment, work is similarly broken up either by product type or production stage. In both cases, the systems being addressed by the IPT should be clearly defined and should be scoped to a manageable size.

3.3. IDENTIFYING CRITICAL SYSTEM REQUIREMENTS

The goal of the VRM methodology is to better deliver what the customer wants. The starting point for variation risk management is identifying the voice of the customer and translating this into the list of critical system requirements.

For those products in development, identifying CSRs is already done as part of the normal product development process. The next three sections should be used to confirm that the current product development process will provide the right information. Additional work may need to be done if product development is lacking a good requirements definition process.

For those products already in production, Sec. 3.3.1 and Sec. 3.3.2 are not relevant and readers can skip to Sec. 3.3.3.

3.3.1. Identify the Voice of the Customer

Early in product development, the product development team members must collect information about what the customer wants. This is critical to the entire VRM methodology. If the wrong customer requirements are identified, then efforts to design a new product or improve an existing product for the customer will be misguided. In some organizations, identifying customer needs is done by special teams as part of the formal product development process; in others, it is the responsibility of the design team. There are many books dedicated to how to define customer requirements for a product.

The first step in defining requirements is to listen to the *voice of the customer.* Customers can include the people purchasing, repairing, or manufacturing the product. For example, customer requirements for a child's toy may include safety, usability, and product life. However, for the parent who has to assemble the product at 2:00 A.M. on Christmas morning, easy assembly of parts without excess force or exotic tools is an important customer requirement.

3.3.2. Identify Specifications and Requirements

The voice of the customer is typically in the form of general and nontechnical terms. The next step is to translate these voices into the technical requirements to which the product will be designed. The technical requirements should have clear target values and tolerances. The translation of the voice of the customer into the technical requirements can be aided through using the House of Quality (Cohen, 1995; Hauser and Clausing, 1988; Revelle et al., 1998). Details of the House of Quality (also called Quality Function Deployment) are given in Appendix C.

Part of the requirements development stage is the prioritization of the most critical technical requirements. These can include those requirements that are important to the customer and those that may be difficult to achieve given the current state of technology. In addition, in complex products, the customer requirements are flowed down and allocated to the individual systems. This process is typically done by the system engineering function.

3.3.3. Identify Critical System Requirements

The system of interest may have many requirements, some more critical than others. To reduce the effort needed to identify system KCs, only critical system requirements should be used as an input to the identification procedure.

For a new design, a requirements document and previous designs are used to identify CSRs. In addition, manufacturability requirements, scrap, and rework data on previous products should be consulted to identify additional potential system KCs. It can be more difficult and easier to identify CSRs retroactively on an existing product. The original IPT often is unavailable for consultation, making it necessary to infer the product rationale. However, there is often data about cost, defects, and customer dissatisfaction that can make the identification procedure simpler.

For both new and existing products, a short list of possible CSR candidates should be identified, and then the team should determine if these requirements are likely to be sensitive to variation. For both new and existing designs, CSRs can be identified from a number of sources:

- **Requirements Documents.** One of the first outputs of the initial steps in the product development process is a set of system requirements. Because requirements documents for each system can include hundreds of entries, it would be too time consuming to examine each one to determine whether it is sensitive to variation. Instead, the first step is to prioritize system requirements to create a shorter list of CSR candidates to work from.
- **Interface Requirements.** It is the system engineer's responsibility to allocate requirements to each system such that if CSRs are met, customer requirements will be met as well. In addition, interfaces between systems can impose critical system requirements on the system of interest. Some of these interface requirements may be sensitive to variation. For example, there may be (Ulrich and Eppinger, 1995).
  —**Mating Requirements** that define how two parts or assemblies fit together. For example, the engine mounting on the frame can cause noise if bolts are not properly torqued and they vibrate or come loose.
  —**Energy Requirements,** which can include power into an electronic system. Fluctuations in the power source can damage sensitive electronic components.
  —**Signal Requirements** in electronics that can be impacted by a connector that does not properly mate.
  —**Material Flows** that can include hydraulic fluid used to control the flaps on an aircraft. Debris contamination of this fluid can reduce the life of the valves.
- **Failure Modes.** There may be other CSRs that do not directly stem from requirements set in response to the voice of the customer. A prioritized list of failure modes, produced by a systematic approach such as failure modes and effects analysis (FMEA) (McDermott et al., 1996), is useful in generating CSRs. In addition, the FMEA process, as currently done by many product development organizations, provides a starting point for variation flowdown because it identifies failure modes as well as potential causes. Further details on FMEAs are provided in Appendix C.

- **Manufacturing (or Producibility) Requirements.** CSRs can also be associated with ease of manufacturing. For example, clearances and assembly tolerances may not directly affect the system's performance, but will impact production's ability to easily build the product and ultimately affect total product cost.
- **Customer Complaint and Warranty Data.** Customer-reported defects in previous designs can highlight hidden requirements. Complaints fall into two categories: those that could have been avoided through better control in production and those associated with design flaws that will not go away no matter how tight the control. Only the first should be considered as potential KCs.
- **Quality Plans and Reports.** Quality control requirements are often related to CSRs and to critical parts and manufacturing processes. The team should concentrate on those QC requirements that generate significant rework or scrap or where the cost of excess variation is high, for example safety-related CSRs. For new products, the team can look at quality reports for previous designs.

3.4. IDENTIFYING SYSTEM KEY CHARACTERISTICS

A system KC is a CSR that is sensitive to variation. In the case of a relatively simple product, the product design IPT will start by identifying the product KCs.[1] In the case of a complex product, the system IPT will start with CSRs for their system and identify their system KCs. For example, if a team is working on the door of a car, system KCs being identified are those directly related to the car door. As stated above, system KCs can derive from functional requirements that directly relate to a CSR (e.g., the seal between the glass and the door), interface requirements (e.g., the shape of the door relative to the car body), or manufacturing requirements (e.g., the ability to easily align the door in the assembly fixture).

3.4.1. What Is a System Key Characteristic?

The IPT should apply the following litmus test to determine whether a CSR is a system KC:

> Is the expected variation in the manufacturing process likely to cause a defect in this critical system requirement?

---

[1]Chapter 9 describes how the system engineering function has the responsibility for identifying product KCs and flowing those to the individual system IPTs. Refer to that chapter for a detailed explanation of the identification procedure within the context of the entire product development process.

One pitfall some teams fall into is to use the *best case* for process variation rather than the *expected* variation. The best-case process capability assumes the newest, best-run machine on a good day. The expected variation used to identify KCs is the likely variation that will be seen under normal conditions. It is easy to justify away KCs by saying, "If we build it perfectly, we will not have any variation." However, perfection is unlikely to be the case in most manufacturing operations.

Another way to screen for KCs is to ask the following about a sampling of products

> Are some of these products unacceptable to the customer? If so, how do the bad products differ from the good products and why?

To ensure that system KCs are identified correctly, the IPT should also ask the following questions:

Is the KC Too General?
It is often tempting to identify as system KCs broad requirements such as reliability, safety, manufacturability, and cost. The team should break down broad CSRs into KCs with quantifiable target values and tolerances. In the case of a medical product that goes inside the human body, the general requirement of safety can be broken down into specific requirements, such as:

- Cannot leak more than x mL of fluid
- Cannot have any protrusions greater than x mm
- Cannot introduce contamination into the bloodstream (zero contamination)

During a team discussion, a requirement that is too broad will often generate competing examples as to why the requirement is a KC in one case and not in another. The team should break the general requirement into more specific requirements that can be evaluated with little debate.

Does the KC Have a Quantifiable Tolerance?
Another way to avoid overly general KCs is to ask what the allowable tolerance is. The tolerance specifies the limits at which variation becomes unacceptable. For example, if power output is a requirement, how much variation is allowable before the customer sees degradation in performance? Once the tolerance is established, the team should debate whether the expected variation is likely to exceed this tolerance.

Has Weight Been Identified as a KC?
Weight almost always arises as a potential KC when an IPT is first exposed to the KC methodology. In most cases weight will not be a system KC be-

cause small variations in dimensions will not cause the allowable tolerance on weight to be violated. However, there are cases where weight is a KC. For example, in aerospace applications, every ounce of weight increases the fuel requirement of the aircraft. Small variations in the application of sealant or the use of shims can significantly increase the weight of an aircraft. Another example where weight is critical is in the balancing of satellites, where the symmetry of the weight is critical. To determine if weight is a KC in their system, the team should return to the litmus test (Is the expected variation in the manufacturing process likely to cause a defect in this CSR?).

Has Cost Been Identified as a KC?

It is often tempting to list cost as a KC because units and tolerances are quantifiable, and cost is sensitive to variation in the manufacturing processes. Cost can include rework, scrap, excess labor, and excess material. However, in most cases cost is the result of a violation of a technical requirement: For example, a car door is reworked because the closing force is wrong. The KC should be the closing force, not the rework cost. The team should endeavor to identify those requirements that drive the cost and should use each of those, not cost, as the system KCs.

3.4.2. Examples of System Key Characteristics

There are several examples that will be used throughout this book to demonstrate concepts and methods (Table 3-1). The examples and their system KCs are introduced in this section.

Aircraft Wings

Variation in an aircraft wing can impact both the performance of the aircraft and the cost of producing the aircraft (Fig. 3-7). While there are many system KCs associated with the wing, we will review only three:

1. **Incidence Angle.** Incidence is the angle between the longitudinal axis of the fuselage and the cordline of the wing. If there is a difference between the incidence of the right and left wings, the plane needs to be

Table 3-1. Examples of system KCs

Aircraft Wing	Printed Circuit Board	Optical Collimator	Automotive Door	Medical Product
Incidence angle	Start-up yield	Output evenness	Closing force	Leakage
Steps and gaps	Long-term yield	Power loss	Leakage	Contamination
Shimming	Mean time to failure		Gap shape around door	

F-22 Radar Signature

The F-22 is a fighter jet that has stealth capabilities (i.e., it is difficult to detect by radar). The shape of the aircraft's hull is one of the characteristics affecting its ability to hide from radar. Hull discontinuities such as steps and gaps or deviations from the designed contour increase the possibility of the aircraft being detected by radar.

When the F-22 was in the design stage, an analytical model picked up a problem with the design. The large number of doors, steps, gaps, and mismatched seams meant that the stealth requirements were unlikely to be met. As a result, the group needed to redesign the underside of the aircraft, adding design cost and delaying a critical design review. Some of the changes involved tightening tolerances on tooling and modifying the layouts of doors and drain holes. The redesign to remove the seams and reduce variation ended up costing in the range of $20 million to $25 million and led to an increase in weight. (Fulghum, 1994a, 1994b). Despite the cost increase, it was better that the issue was found earlier than later. Estimates showed that finding the problems in the design stage might have saved tens to hundreds of millions of dollars in repairs and late design changes.

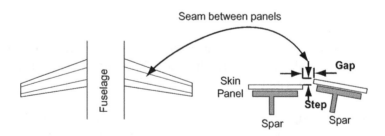

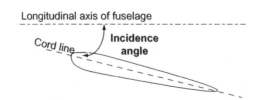

Figure 3-7. System KCs for an aircraft wing.

"trimmed." Trimming increases the overall drag of the aircraft and raises fuel costs.

2. **Steps/Gaps.** The wing has an underlying structure of spars and ribs, and its surface is made of panels of aluminum. Where the skin panels meet, there is always a gap and there may be a step between the two. Steps and gaps increase turbulence which increases drag and fuel consumption.

3. **Shims.** When the wing is attached to the body of the aircraft, the fit between the wing stub (the structure at the end of the wing) and the centerbox (the rigid structure in the center of the fuselage that supports the wing) must be exact. Because both the wing stub and the centerbox are assembled structures, there can be variation in both. As a result, shims may be required to make up the gaps between the two structures. If variation is significant enough, tapered shims may need to be custom-built. Fabricating custom shims increases the manufacturing cycle time, and shims add to the weight of the aircraft.

Printed Circuit Boards (PCBs)

PCBs are used in many products today, from phones to portable stereos to children's toys. The typical system KCs for printed circuit boards include:

- **Yield.** Variation in the chip placement, lead size and placement, and solder quality can result in a board failing initial testing. The cost of poor yields can include scrapping the boards and reworking the boards. Yields vary considerably with time. Often start-up yield is poor; then, as manufacturing processes are tuned, yields improve, reaching a steady-state performance.

- **Mean Time to Failure (MTTF).** Often PCBs will work when they are first produced but may fail after some time of use. The MTTF is one measure of reliability. If the board fails within the warranty period of the product, the manufacturer may incur significant cost to either fix or replace the defective product. The time to failure can be a function of weaknesses in the solder joints leading to open circuits, contamination causing shorting between runs, or failure of one of the components. If the board fails outside the warranty period it can lead to customer dissatisfaction and lost sales.

Optical Collimators

A collimator takes a light source, such as a laser, and creates a focused beam with a uniform intensity across the diameter of the beam (Fig. 3-8). Collimators are used in many applications from telescopes to medical lasers. The

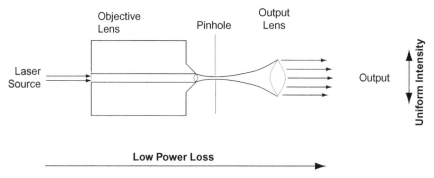

Figure 3-8. System KCs for a collimator.

unmodified output of a laser can be very uneven if targeted directly on a surface, and there can be increased dispersion depending on the distance between the laser and the object being illuminated.

The first lens (the objective lens) focuses the light source. The shape of the output beam has a characteristic "waist" that represents a narrow point in the beam. A pinhole aligned with this waist has the effect of removing much of the random noise in the light source. The light is then sent through an output lens that creates a parallel, focused and even intensity beam of the desired diameter. There are many requirements of a collimator, the two primary ones being a low power loss through the system and a uniform intensity across the diameter. Power loss is important because a higher-power laser may be required, increasing costs. The output beam must have an even power distribution. Especially in medical applications, it is detrimental to have "hot" and "cold" spots in the output.

The performance of optical systems like collimators is difficult to model using closed form solutions. Rather, simulation tools are used to do ray-tracing diffraction, color correction, and tolerancing.

Disposable Medical Products

There is a class of medical products that either collect fluids from or dispense fluids to a patient. These can include blood transfusion kits, catheters, dialysis kits, intravenous (IV) kits, and pumps. These products are commonly disposable; they are used once and then thrown away. Common to all products are two system KCs: leaks and contamination.

- **Leaks.** Many disposables are made up of tubing, bags, cannulas, needles, filters, and other plastic devices. Components are joined together using welding, thermal melting, adhesives, or solvents. These assembly processes can be difficult to control and can result in small defects that permit leaks. Leaks can make a product nonsterile and/or cause a biohazard.

- **Contamination.** Although most disposables are manufactured in clean rooms, it is possible for contamination (dirt, debris from assembly, dust, etc.) to get caught in the fluid path of the product. Contamination can be dangerous if it gets transferred to a patient or blocks a critical fluid flow. In addition, contamination can occur outside the fluid path causing sterility issues. Manufacturers of disposables spend significant time controlling processes and inspecting to ensure that no contamination is included in their products.

Noise Vibration and Harshness (NVH)

NVH is a collection of negative impacts on automobile passengers including rattles, wind noise, and excess vibration, that would make the passenger view the car as being of inferior quality. In the last decade, the bar has been significantly raised on NVH as quality has become as important as cost, power, styling, and the newest bells and whistles. Most automotive design divisions have acoustics experts on staff whose sole job is to quantify, diagnose, and reduce the sources and impact of NVH. NVH characteristics are measured in terms of both decibel level and frequency of the noise or vibration.

Many parts of a car can contribute to NVH and the passenger's experience, including the frame; seals; heating, ventilation, and air conditioning (HVAC) system; and brakes. In addition, noises are additive and can be either amplified or dampened by various systems. NVH can be reduced in several ways including:

1. **Reduce the Source of Noise.** Sources of noise can be removed by a variety of means including stiffening parts, reducing errors in assembly, and tuning parts to the right natural frequencies.

2. **Reduce the Impact of Noise on the Passenger.** Two methods of reducing the impact of noise include sound absorbing insulation and employing active noise cancellation technology. Both of these methods can increase the cost and the weight of a car as well as the complexity of the assembly process.

The challenge of NVH is that it is very difficult to quantify and model a priori in product development. Often it is necessary to make qualitative estimates about sources and their relative contributions to the passenger experience. However, the ability to quantitatively model NVH is constantly improving.

3.5. CREATING THE VARIATION FLOWDOWN

Once system KCs have been identified and quantified, a variation flowdown should be methodically generated by the IPT. The process for creating flow-downs for products under development is virtually the same as that for existing products. The major differences are that information sources differ and there is more certainty about which part and process KCs impact system KCs for exist-ing products. The variation flowdown creates a picture of how the quality in a system KC is delivered by the assemblies, individual parts, manufacturing processes, and suppliers. A number of methods exist for creating and recording the variation flowdowns including House of Quality, FMEA, and fishbone dia-grams, all of which are reviewed later. The method here has been found to be broadly useful and avoids many of the shortcomings of other methods.

Figures 3-9 and 3-10 show two examples of variation flowdowns for the collimator and disposable assembly described in the previous section. Two different formats, vertical and horizontal, are shown. In the first, the power loss and all of its contributors are listed. The second, shows the flowdowns for three system KCs.

It is tempting for teams to jump from identifying critical system require-ments to identifying the individual part and process KCs that should be con-trolled. An example illustrated in Fig. 3-11 shows two methods for attaching

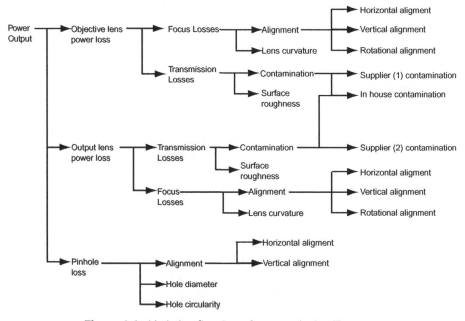

Figure 3-9. Variation flowdown for an optical collimator.

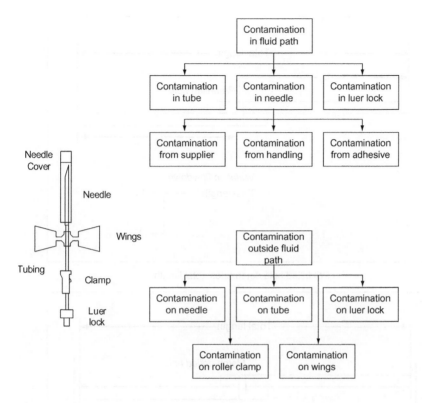

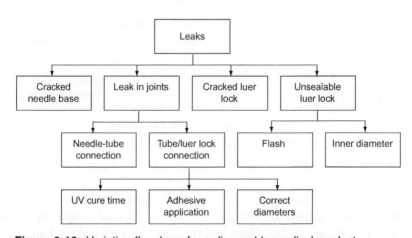

Figure 3-10. Variation flowdown for a disposable medical product.

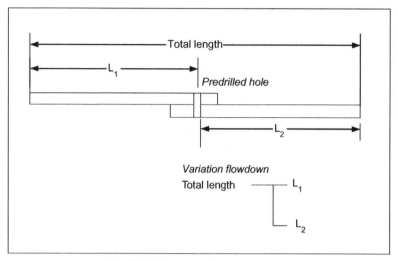

Predrilled hole determines total length

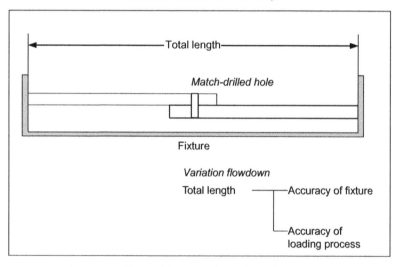

Fixture determines the total length

Figure 3-11. Different assembly techniques give different flowdowns.

two panels. In the first method, which uses a predrilled hole, the lengths from ends to holes drilled in each part determine the total length. In the second method, a fixture locates parts, and the parts are match-drilled and fastened. In this method the total length is a function of the accuracy of the fixture and how well the part is loaded into the fixture. Team members at one organization failed to analyze the assembly process when defining the KCs, and, although they were using the fixture approach to assembly, they had unnecessarily specified the end-to-hole dimensions as critical. This resulted in unnecessary control being applied to a dimension that had no impact on the final product quality.

KCs should be identified methodically using either a top-down or bottom-up process or both. Ideally, the team should go through both, starting with the top-down process and using the bottom-up process as a check on the results. A variation flowdown may have as few as 2 or as many as 10 layers. Any system with more than 10 layers should be broken into smaller systems to reduce complexity.

In addition to providing the framework for the assessment phase, the variation flowdown has a number of important uses.

- **Evaluation of the Impact of Changes.** In addition to manufacturing variation, a major source of quality issues is the impact of changes in parts and manufacturing processes after production is under way. For example, the change in inspection frequency was found to be a possible contributor to the Alaska Airlines Flight 261 crash (NTSB, 2002).

  While not all future consequences can be identified, some may be avoided. When a change is requested, the variation flowdown should be consulted to understand which system KCs are likely to be impacted. The variation flowdown should then be updated as necessary.

- **Root Cause Analysis.** When unexpected quality degradation occurs during production, significant time is often spent understanding the root cause. The engineering intent of the designers must often be inferred by those tasked with increasing yields, either because the original designers are not available or because time has elapsed and people have forgotten the exact design rationale. The variation flowdown can provide a record of the design team's thoughts and rationale relating to quality and may provide some insight when facing a variation issue during production.

- **Identification of Critical Processes.** While a typical variation flowdown will look like a pyramid (few system KCs at the top and many part and process KCs at the bottom), in some cases it can look like a diamond (Fig. 3-12). This shape occurs when a single process has an impact on many parent KCs. For example, in one medical product a number of seemingly random failures were occurring. The failures seemed to come and go without cause or pattern, but as a group they represented a significant portion of the warranty issues. Some of the failures were shorts in the electrical system, some were leaks, and others were cosmetic damage. When the IPT created a variation flowdown for the product, it became immediately apparent that the key issue was the cleaning process. The cleaning process and chemicals used by the customer were found to be possible contributors to each complaint type. The negative

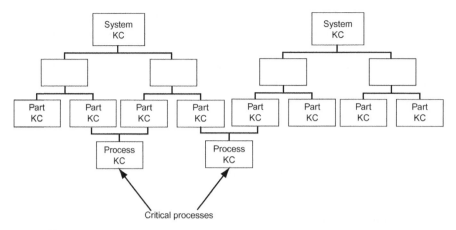

Figure 3-12. Using a variation flowdown to identify critical processes.

impact of cleaning on each individual failure was small. However, when the impact of the cleaning process was summed across all possible failures it could cause, it turned out to be significant. Changing the cleaning process significantly reduced the total failure rate. Identifying the process KCs that impact many part and system KCs can be of significant benefit to the product. It's like a two-for-one sale: By controlling a single process, the benefits to the product are multiplied. The return on investment for fixing these critical processes is significant.

3.5.1. What Information to Gather

The processes for creating variation flowdowns for existing products are the same as those used to create variation flowdowns in product development. Variation flowdowns for existing products can often be created by a smaller team than for new products because much of the information required has been documented. In both cases teams should gather the following information before proceeding with a variation flowdown.

- **System KCs.** These are the starting points. In addition to performance requirement, system KCs can include interface requirements, failure prevention requirements, and manufacturing requirements.
- **Product Architecture.** The major systems, assemblies, and parts should be identified. The flowdown will parallel the product architecture.
- **Functional Structure.** This provides an understanding of how the product or system will function and how the major systems will interact.
- **Manufacturing Processes.** The team should understand how the product

will be built in order to identify the process KCs. Spending time tracing how parts go through the production process is a good way to understand how the product is assembled, where variation is introduced, and how variation in part and process KCs will ultimately impact system KCs.

- **Critical Suppliers.** Parts and materials supplied by outside vendors that impact product quality must be identified. The capabilities of suppliers must be understood. In some cases it may be necessary to work with suppliers to achieve the appropriate levels of quality. Roles and responsibilities of the supplier base are outlined in Chap. 10.

- **Previous or Current Products.** When the team is discussing possible contributors to variation, it is a good idea to use examples from previous designs. Often, the manufacturing and quality groups have a strong understanding of issues likely to resurface in new designs.

- **Quality Control Plans and Data.** Current quality control plans and quality data are good sources of information about how variation is being controlled for current or similar products. These sources of information can indicate which KCs are important or have been problematic in the past. The team should review:

  —**Scrap/Rework Data.** The team should identify the major sources of scrap and rework and should understand their root causes.

How Many KCs?

"How many KCs should I identify?" is a question that arises frequently. It reminds me of assigning reports while teaching. Students always ask, "How long does it have to be?" Every time a minimum page limit was set, the weakest students would fill the report with non-relevant material and the overeager students would hand me reports longer than most Ph.D. theses.

The answer is, it depends. You need to create a variation flowdown that is comprehensive enough to capture all the major contributors to system KC defects. You will not take action on or evaluate every single KC. In general, use the reasonableness test—do not continue to flow down branches where the contribution to variation is minimal.

A typical system KC will flow down four to five layers and each branch will have three to five child KCs. However, this should not be used as a hard and fast rule.

—**FMEAs.** If FMEAs have been done, the reports may contain valuable information about failure modes and causes.

—**Customer Complaint and Warranty Data.** This data is useful if the data points to failures in production.

—**SPC Data.** This data will indicate what processes are critical to quality.

3.5.2. How to Conduct the Top-Down Process

Figure 3-13 shows the top-down process. The top-down process starts with system KCs and identifies how each assembly contributes to the system variation. Once the assembly KCs are identified, the team then looks at what part and process KCs contribute to each of the assembly KCs and so forth until a supplied part process is identified. The advantage of the top-down approach is that at each layer the team can determine if the right KCs are identified. In addition, the top-down process ensures that every part and process KC can be traced to a customer requirement.

Two questions should be asked each time a KC is suggested:

1. **What Are the Units and Values for the Target and Tolerance?** This question ensures that the KCs will be specific. It also forces the IPT members to begin to quantify the acceptable amount of variation.

2. **Why Is This KC Important, and Is It Likely to Contribute Significantly to System Defects?** There should be a justification for each KC. This prevents issues from being added that:

   —Could be a problem only on a rare occasion

   —Are not sensitive to variation

   —Are a someone's pet issues

Even if the contribution is small, make sure the KC is documented. Make a note that the KC's contribution is small and stop the flowdown on this branch of the tree. It is important to record the small contributors because at a later date it may become clear that assumptions were inaccurate and more detail is required.

The top-down process can be aided by a number of tools. For each KC, the team should ask, "Is there a function or tolerance chain that describes how the KC is created?" If there is, evaluate the function or tolerance chain and identify all the contributing KCs. If there is none, the team should analyze the product by looking at each assembly, part, and process and identifying what dimensions and features may impact the system KC.

After the first layer of the flowdown is completed, the same process should be reiterated for each assembly KC to determine its child KCs. The flowdown process is complete when a manufacturing process or supplied

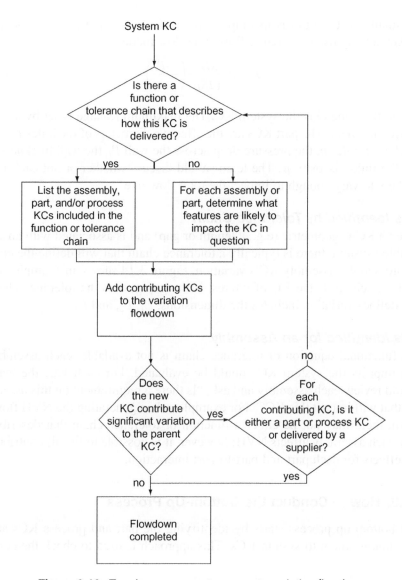

Figure 3-13. Top-down process to generate variation flowdown.

part is reached. Suppliers can be asked to complete the flowdowns for their supplied parts (see Chap. 10).

KCs Identified by Function

When a KC is nongeometric (i.e., force, voltage, and so on), there is often a mathematical function that describes how the KC is delivered. For example, a DC voltage and a resistance will determine a DC current. Another example is the leakage of fluid through small defects in a wall. Assuming the defects are small and the fluid is incompressible and viscous, the flow can be approximated by using a circular Poiseuille flow model

$$Q = n \frac{\pi D^4}{128\mu} \left(-\frac{P_i}{t} \right) \qquad (3\text{-}1)$$

The flow rate Q is the system KC. The maximum leakage is set by a system requirement. The part KCs are the effective diameters of the holes D, the number of holes n, the pressure drop across the wall P_i, the wall thickness t, and the fluid viscosity μ. The team should discuss whether or not each KC is likely to vary enough to impact the total flow rate.

KCs Identified by Tolerance Chain

When a KC is geometric (e.g., a length or gap) and is associated with an assembled system, there is typically a tolerance chain that will define the contributors to the assembly KC's variation. Figure 3-14 shows an example of a tolerance chain. If the KC of interest is the gap g, then the tolerance chain that defines variation includes the dimensions, L_1, L_2, and L_3.

KCs Identified for an Assembly

If a functional equation or tolerance chain is not available, each assembly that impacts the system KC should be evaluated. For each KC, the team should review each assembly and ask, "Is there a requirement on this assembly that will impact this KC?" For example, when generating the NVH flowdown, there may not be either a function or tolerance chain that describes how each assembly affects NVH; however, it is possible to list all contributing effects for each part and part-to-part interaction.

3.5.3. How to Conduct the Bottom-Up Process

The bottom-up process starts by identifying the part and process KCs and then linking them to system KCs. This approach is used to check the com-

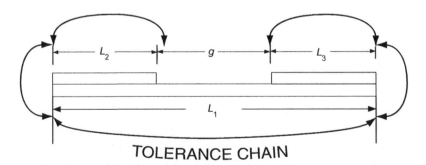

Figure 3-14. Tolerance chain example.

pleteness of the flowdown. In many cases the engineers have already identified the critical tolerances on individual parts based on engineering judgment
(Fig. 3-15). As with the top-down process, the team should ask, What are the
units and values for the target and tolerance? and Why is this KC important,
and is it likely to contribute significantly to system defects?

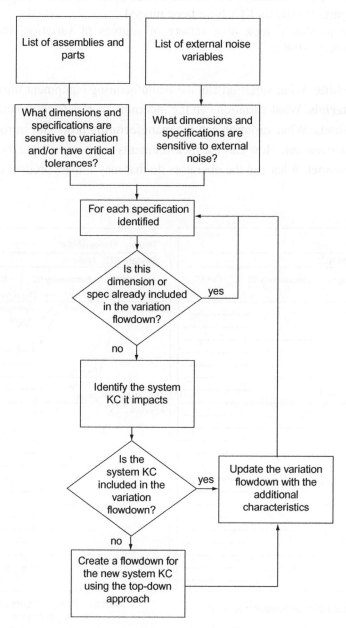

Figure 3-15. Bottom-up process for generating variation flowdowns.

In the bottom-up process, go through each assembly, part, manufacturing process, and interface and ask, "What tolerances are important and what processes may be incapable?" Then ask why. Each part or process KC identified should be linked to a system KC. If critical dimensions or requirements are identified and cannot be tied to an existing system KC, the system KC list should be reevaluated to ensure that it is complete and that no system, assembly, part, or process KCs have been missed.

The team should look at a variety of sources of variation, including (Montgomery, 1996):

- **Machine.** What variation can the manufacturing equipment introduce?
- **Materials.** What variation can the incoming materials contribute?
- **Methods.** What variation can the manufacturing processes introduce?
- **Measurement.** How can the measurements error impact quality?
- **Personnel.** What can the operators do that may impact product quality?

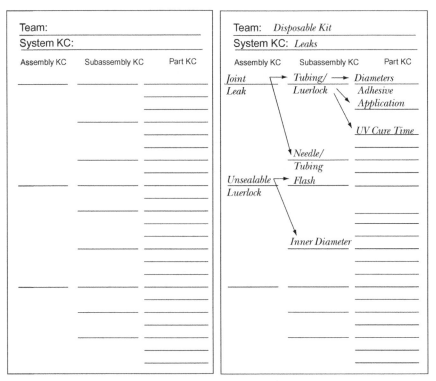

Blank variation flowdown template

Variation flowdown template with sample entries

Figure 3-16. Template for recording variation flowdown.

Previous designs also provide a starting point for the bottom-up process. The critical tolerances and manufacturing processes should be identified and evaluated in the context of the new design. However, just because a part or manufacturing process was considered critical on a previous design does not mean that it should automatically be considered a KC in the new design. Changes to configuration, production processes, and requirements should be assessed to determine if the part or manufacturing process KC is critical on the new design.

3.5.4. How to Conduct and Document the Identification Procedure

The variation flowdown is typically generated in team meetings at which all IPT members actively participate. Ideally, the flowdown should be generated

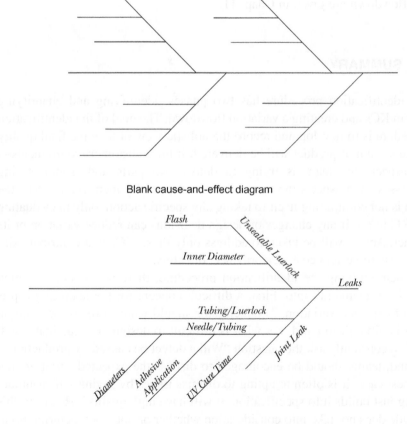

Blank cause-and-effect diagram

Sample filled-out cause-and-effect diagram

Figure 3-17. Cause-and-effect diagram.

with the aid of a facilitator who can lead the group and keep it focused. The IPT members should be asked to bring their relevant data and documents to the meeting, including all of the information listed earlier in this chapter. The first step is to generate the list of system KCs. Once this is done, a variation flowdown for each system KC should be generated. A facilitator should be responsible assigning a person to act as a recorder for documenting the results of the process. Documentation can be done in one of two ways. The first is to use a software system such as Visio or PowerPoint to record the data in real time and project it back to the team using an LCD projector. If the keyboarding skills of the recorder are insufficient or the equipment is not available, use a paper form for recording the information and then transfer the information to electronic format at the end of the meeting.

Figure 3-16 shows a sample template that can be used to aid in the flowdown process. Alternatively, a more traditional cause-and-effect diagram (also called a fishbone diagram) can be used to organize the flowdown (Fig. 3-17). Additional methods for documenting the system KCs and variation flowdown are given in Chap. 11.

3.6. SUMMARY

The identification procedure has two phases: identifying and quantifying system KCs and creating a variation flowdown. The goal of the identification procedure is to develop and record the holistic view of how the final quality of the system or product will be delivered. It links customer requirements— the aspects the team is trying to deliver—to parts and manufacturing processes—the aspects the team has control over. By identifying a KC, the team is not committing itself to taking any specific action, only to evaluating the KC to see if any changes or design decisions can reduce variation or its impact. Action will be taken to address only those KCs that contribute significantly to system cost, performance, or safety.

When applying the identification procedure, there are several common pitfalls all teams fall into. First, a difficult concept for the team to grasp is that of expected variation. The facilitator should work to focus the team on what is difficult to build, not what is difficult to design. The facilitator will need to constantly ask the question "Will a defect be caused by production?" Second, teams should be encouraged to discuss the expected variation, not the best case. It is often tempting to dismiss issues by saying, "If manufacturing just builds it to specification, it will have high quality." However, this attitude does not take into consideration whether or not manufacturing is capable of building it to specification at a reasonable cost.

Third, teams often get bogged down in minute details. A flowdown for a single system KC should take about an hour to draft. It is important for the facilitator to prevent the meetings from digressing into design or engineering reviews. This is not a time to be redesigning the product, but to be discussing the current state of the design and where it is most sensitive to variation.

Figure 3-18 shows a summary of the identification procedure. The IPT should start by defining the scope of their VRM efforts. Next, the IPT should collect the relevant data to support the identification procedure. Sources of information include the following:

- **Product Architecture.** The major assemblies and parts and their relationships define the systems addressed by each IPT.
- **Requirements (System, Interface, and Manufacturing).** These are the starting points for identifying critical system requirements, which are analyzed to determine system KCs.
- **FMEA.** Failure modes and effects analysis tables can also be a good source for system KCs as well as an aid in the flowdown process.
- **Manufacturing Processes.** The way how the product will be manufactured is used to identify the process KCs.
- **Customer Complaint and Warranty Data.** The customer complaint data includes information about warranty data, returns, and repairs. This data can help identify where the manufacturing processes are not delivering products with adequate quality.
- **Quality Plans and Reports.** For existing products it is possible to identify what parts and manufacturing processes are currently failing to achieve the quality targets or what was considered critical on previous designs. These plans and reports can include SPC plans, inspection and testing procedures, and scrap and rework records.
- **Process Capability.** The existing or surrogate process capability will help determine whether or not the manufacturing process is capable of producing system KCs at a high enough quality. At this stage, teams should use qualitative determinations of capability. The assessment phase will use quantitative data.
- **System KC Tolerances.** In order to determine if CSRs are KCs, the team needs to understand what the target tolerances are. These are compared to process capability to make a preliminary determination that the CSR is likely to be impacted by variation.
- **Functional Structure.** The way the product function delivers system KCs is used in the top-down variation flowdown process. The functional structure defines how the part, assembly, and process KCs combine to deliver system KCs.

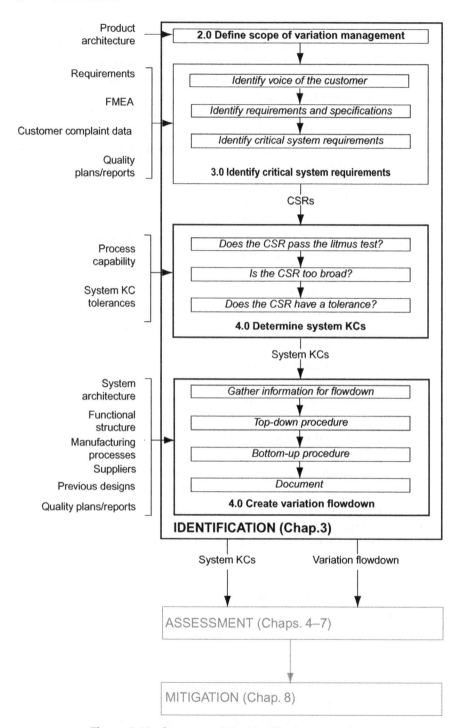

Figure 3-18. Summary of the identification procedure.

- **Suppliers.** Some of the KCs will apply to materials and parts delivered by outside vendors.
- **Previous Designs.** Information about the performance, manufacturability, and warranty performance of similar products can be used to identify potential weaknesses in the new product.

Once the data is collected, an IPT should identify the system KCs and then create the variation flowdown using both the top-down and bottom-up approaches.

- **Identify Critical System Requirements.** From the voice of the customer, the team needs to narrow requirements down to a small subset that are most important (i.e., the CSRs). This will reduce the overhead in the system KC identification procedure. The CSRs include the performance, interface and manufacturability requirements.
- **Identify System KCs.** For each CSR, the team should ask the following questions to determine whether it is a system KC:
  - —**Does the CSR Pass the Litmus Test?** If the CSR is likely to be negatively impacted by variation in the production process or by vendor-supplied materials or parts, then it should be included in the list of system KCs.
  - —**Is the CSR Too Broad?** As pointed out earlier, some CSRs are too general (reliability, safety, etc.). The team should break down general CSRs into specific requirements.
  - —**Does the CSR Have a Target Value and a Tolerance?** Another tool to ensure that the team is being specific enough is to make sure the CSRs being addressed have unit values, a target value, and an allowable tolerance.
- **Create Variation Flowdown.** Finally the variation flowdown should be created. The team should continue to flow down the KCs until either a single part or manufacturing process at the end of a branch is reached. If a KC for a supplied part or a KC with only a minimal contribution to system variation is encountered, the branch of the flowdown should be terminated. The flowdown should be documented for use in the assessment and mitigation phases.

The outputs of the identification procedure are system KCs and a variation flowdown containing the assembly, part, and process KCs. The next step is to assess the relative probability of exceeding the allowable tolerance on each system KC and the cost of defects. The next four chapters describe the assessment procedure.

4

OVERVIEW OF ASSESSMENT

This chapter reviews assessment, the second step in the VRM Methodology. Chapters 5 through 7 contain further details of the steps within the assessment procedure. All readers should review this chapter.

The identification procedure results in two outputs: system KCs and a variation flowdown. System KCs are variation-sensitive critical system requirements, and a variation flowdown is a hierarchy of dimensions, features, and parameters that might—based on engineering judgment—impact the product defect rate. Mitigation (Chap. 8) reduces the sources and/or the impacts of variation. The assessment procedure is important because it is impossible to apply mitigation strategies to all identified KCs. The middle step, assessment, prioritizes the KCs using quantitative and qualitative methods. Assessment has two outputs:

1. **The Risk Associated with Each System KC.** Risk is a combined measure of the expected probability and cost of a defect. Reducing the probability and cost of defects in system KCs is what ultimately matters to the product development and production communities. While attaining the assigned tolerances on individual parts is important, it is only important because it drives the ability to achieve system requirements.

   —The Probability of Exceeding the Allowable Tolerance in System KCs. In the case of new designs, the probability must be predicted; for products in production, the actual defect rate can be measured based on the scrap and rework data.

—**The Cost of Missing the Allowable Tolerance.** The team must take into consideration the cost of a defect when prioritizing their efforts. Costs can include warranty, scrap, rework, lost capacity, and inventory. By using defect cost as the single metric for each assembly KC, part KC, process KC, and system KC and summing costs up through the system level, the total cost impact of variation on the system can be determined.

2. **The Relative Contribution of Each Part KC to Variation of the System KC.** Each part and process KC will contribute a different amount of variation to system KCs. It is important to identify which part and process KCs contribute the most in order to determine where mitigation strategies will have the largest potential impact. For example, if one part KC contributes 10 percent of variation, and another contributes 90 percent, the team is likely to get more significant results by addressing the 90 percent contributor.

As with the identification phase, assessment is executed by the integrated product team. Assessment is applied to both products in product development as well as those already in production.

- **Assessment during Product Development.** The IPT should predict both the expected defect rate of system KCs and which part and process KCs contribute the most to variation in system KCs. In addition, the IPT should predict the cost of defects. System KCs are ranked based on risk (the combined expected cost and expected defect rate). Because the design has not been built yet, the team must rely on models, historical data, and prototypes to predict the relative risk.

- **Assessment during Production.** During production, the actual defect rates and cost of defects can be determined. In production, assessment collects all data about cost and defect rates to compute the total cost of variation for the product. In addition, it is necessary to determine how well the current quality control plan will enable the team to both detect and diagnose the defects in system KCs.

Assessment supports variation risk management in two ways. First, assessment provides a data-driven approach for selecting the right problems to work on. By quantitatively modeling or measuring the cost and probability of variation, the team can determine what parts of the design are either at risk or currently costing the organization the most. The assessment is based on historical process capability, models of variation, and production data. For products being manufactured, the assessment uses data from the shop floor and multiple functional groups to quantify the current total cost of quality.

Second, the assessment is a holistic approach that measures the cost and impact of variation on the entire organization, not just on one functional group. This insures optimal solutions for the whole company not just one functional group. The holistic approach is used both for products being designed and being manufactured.

For new designs, there are two assessment approaches: qualitative and quantitative. Qualitative approaches rank the relative cost, probability and contribution based on engineering experience. In the early stages of product development, qualitative assessment is performed in order to identify and rank areas require more detailed analysis. As the product development progresses, and the high-risk and high-uncertainty areas are identified, more precise quantitative modeling tools are applied. Using a qualitative approach first and a quantitative approach later means the total assessment time is less and uses the product development team's limited resources efficiently.

The qualitative assessment is typically done in a group setting with the IPT drawing on the expertise of all team members to determine the relative ranking of system KCs and to estimate the contribution of part and process KCs to system KC risks. The quantitative assessment is typically performed outside of the IPT meetings because it is too time consuming to do the analysis in meetings. Typically, the team will divide up responsibility for assessing various system KCs and the individuals responsible will report the results to the entire group. Based on the assessment results, the team can plan the best mitigation strategy (Chap. 8). For products in production, assessment is based on actual costs incurred when due.

This chapter provides an overview of assessment as applied to new products during product development and to existing products being manufactured. The individual steps in assessment are detailed in Chaps. 5, 6, and 7.

- Chapter 5 reviews tools used to predict and measure the probability or frequency of system defects and tools used to estimate the contributions of part and process KCs.
- Chapter 6 reviews tools used to estimate and measure the cost of system variation.
- Chapter 7 reviews tools used to evaluate the effectiveness of the current quality control systems for products in production.

4.1. ASSESSMENT DURING PRODUCT DEVELOPMENT

The goal during the development of a new product is to identify where the design is incapable of delivering the required quality given the current process capability. Based on the assessment, the team may choose one of a number of possible mitigation strategies (Chap. 8). The goal of assessment during product development is to prioritize which system KCs are at the highest risk. To reiterate, only system KCs have risk associated with them. The part and process KCs only contribute to that risk and are the targets of mitigation. For example, a car's owner does not care if the door perimeter varies more than specified, only that the wind noise is minimal and that it looks good. Reducing variation in the contributing KCs reduces variation in the parent system KC.

The point of the assessment phase is to develop a relative measure of risk and contribution, not to predict the yields and costs exactly. Most engineers are uncomfortable committing to exact numbers for yield and cost, especially early in product development.

The team should group system KCs into four "buckets" (Fig. 4-1). This approach to ranking serves to identify the system KCs that require immediate attention and separate them from those that can wait. The four buckets are high risk, medium risk, low risk, and uncertain risk. The last bucket is for those system KCs for which sufficient information about the design and/or the manufacturing processes is not available to rank the risks. This is especially true when there is new technology in a product or manufacturing process.

Figure 4-2 shows how assessment is used during product development. The inputs from the identification phase are:

1. **Variation Flowdown.** The variation flowdown provides the framework for creating a variation model.
2. **System KCs.** For each system KC, the allowable tolerance should have been specified in the identification phase. Tolerances, along with the variation model and process capability data, are used to determine if the current manufacturing processes are capable of achieving target tolerances.

There are four steps in the assessment procedure:

1. **Assess the Frequency of and Contribution to Defects (Chap. 5).** Using qualitative or quantitative models, the team should assess the

Figure 4-1. System KCs are bucketed according to relative risk.

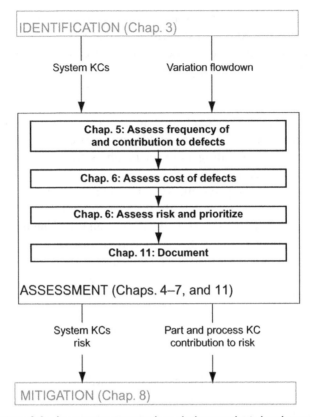

Figure 4-2. Assessment procedure during product development.

frequency of defects in system KCs and identify which part and process KCs contribute the most to variation in system KCs.

— **Compute the Probability of a Defect in Each System KC (Chap. 5).** This can be done either qualitatively on a scale of 1 to 10 or quantitatively using the methods described in the next chapter. The ability to use a quantitative model is dependent on the existence of a variation model and good process capability data.

— **Identify the Contributions of Part and Process KCs (Chap. 5).** Using either qualitative or quantitative methods, the relative contribution of the part and process KCs should be evaluated. This can be done based on the sensitivities and the part and process KC variation, the results of Monte Carlo simulation, estimates, or engineering judgment.

2. **Compute the Cost of a Defect in Each System KC (Chap. 6).** This can be done either qualitatively on a scale of 1 to 10 or quantitatively

using the methods described in Chapter 6. The cost is a function of many factors including the cost of scraping or reworking a part or assembly, or of a defect escaping to the customer. In addition, defects can impact production capacity or inventory costs.

3. **Compute the Risks and Bucket the System KCs.** Based on the cost and probability of a defect, the team needs to compute the risk and bucket system KCs into high, medium, low and uncertain categories.

4. **Document the Results (Chap. 11).** The team must always document the results of the assessment process. In Chap. 11, a suggested format for documenting the results for the entire I-A-M process is given.

4.2. ASSESSMENT DURING PRODUCTION

For existing products being manufactured, more concrete data exists. Current variation costs and defect rates are often tracked and recorded and can be used to compute the real total cost of variation. The team must collect the data and compute the total cost of variation and the relative contributions of the part and process KCs to the total cost. In addition, current quality control methods—including inspection, SPC, and testing, should be evaluated for their effectiveness in identifying changes in process capability and containing defects before they escape to the customer.

Figure 4-3 shows the steps in assessing the impact of variation in a product already being manufactured.

- **Measure the Defect Rates (Chap. 5).** Using a multitude of data sources, the defect rates (both internally or externally detected) can be usually measured.

- **Compute Cost Impact for Each Source of Variation (Chap. 6).** This can be done either by rolling up costs to the system KC level or by identifying the cost associated with each part and process KC. Based on this analysis, the highest-priority areas can be identified. In addition, the contribution of each part and process KC can be measured, computed, or estimated.

- **Create a Scheme for Recording Data (Chap. 6).** Based on the variation flowdown, the team should develop a method for recording costs of defects in the various KCs in the product. A variation flowdown represented in a table format is one way of recording the data and allocating costs to the various KCs.

- **Analyze the Effectiveness of the Current Quality Controls System (Chap. 7).** Using scatter diagram and the quality effectiveness matrix

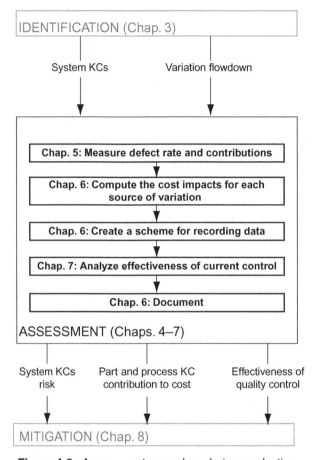

Figure 4-3. Assessment procedure during production.

the effectiveness of the current controls can be determined and possible improvements identified.
- **Document (Chap. 6).** Using the documentation methods described in Chaps. 6 and 7, the results should be documented.

After assessment is completed, the team will have the following information:

- **System KCs Risk.** System KCs are ranked into collections of high, medium, low, and uncertain risk. Once the key characteristics are ranked, a reality check based on the engineering judgment of the IPT members should be applied to the results—"Are there items that have been ranked high risk that are not that risky, or are there items that are low risk that should be ranked higher?" If discrepancies exist, the team should reevaluate the risk numbers.

- **Part and Process KC Contribution to Risk.** The part and process KCs that are large contributors to each system KC risk are identified. These can include individual part dimensions, critical suppliers, and manufacturing processes.
- **Effectiveness of Current Quality Control.** QC that is either ineffective (not capturing defects) or inefficient (too expensive for the return) has also been identified.

5

ASSESSMENT OF DEFECT RATES

This chapter presents models for assessing the probabilities of system defects and estimating the contributions of part and process KCs to overall system risks. The material in this chapter is fairly detailed. It should be used as a reference for the team members responsible for assessment and can be skimmed by others.

During product development, the goal of assessment is to *predict,* based on a model of the design and information about process capability, the probability of a defect for each system KC and determine the major contributors to the defect rate. For designs in product development, the team will need to estimate the frequency of defects and the contribution based on quantitative and qualitative models. For products already in production, the team should use existing data on the defect rates and process capability to do the analysis. Figures 5-1 and 5-2 give the detailed steps for both scenarios. The next four sections detail the tools and methods for each of the steps.

5.1. PREDICTING THE FREQUENCY OF DEFECTS

The probability of creating a defect is a function of two factors: the expected probability distribution of the system KC and its tolerance limits. The expected distribution is the result of two factors: the design and expected process capability. The design will determine how sensitive system KCs are to variation introduced by manufacturing processes, suppliers, and external noise factors. Because the actual product is not available until late in product development, a variation model and variation simulation tools are used to predict system performance. The expected process capability is based on previous performance of similar processes (Appendix B).

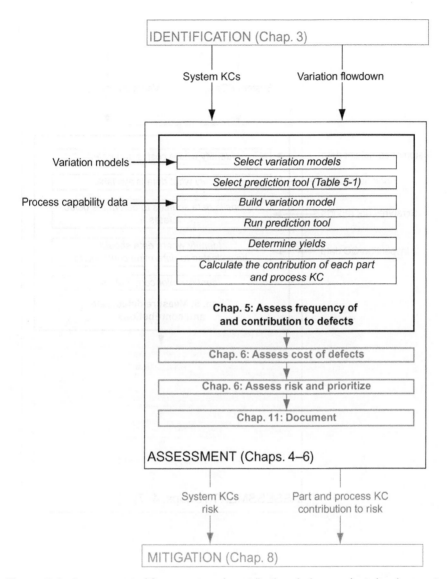

Figure 5-1. Assessment of frequency and contribution during product development.

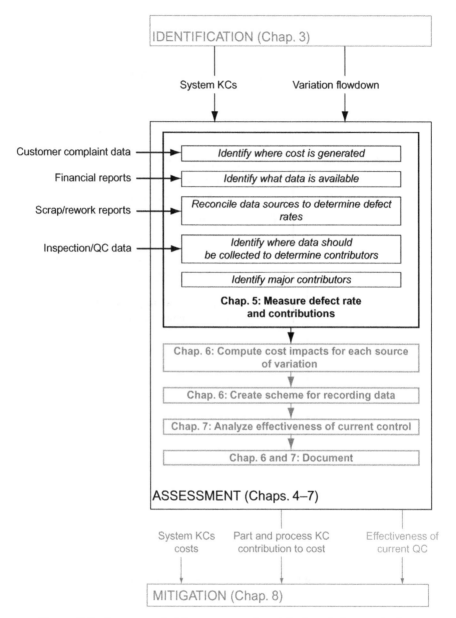

Figure 5-2. Assessment of frequency and contribution during production.

Many products have target specifications that will always be achieved (i.e., a defective product will not be passed to a customer) but at a potentially high cost. For example, a satellite is unlikely to be sent into space unbalanced. However, significant time and money may be spent bringing the system into specification. When assessing the probability of a defect, the IPT should determine the number of systems requiring rework or significant tuning, not the number that will ultimately be delivered out of specification.

To predict variation in system KCs, two decisions need to be made:

1. How to model manufacturing variation and its impact on the system KCs
2. How to predict system KCs variation using the model

Teams should determine what model to use based on the following questions:

- **What Models Are Available?** If quantitative performance models already exist, then the team should make use of them. These can include simulations, back-of-the-envelope calculations, and FEA models.
- **How Accurate Is Each Model?** Some models will give better predictions of the final defect rate than others.
- **How Much Time Will the Assessment Take?** Perfection is the enemy of completion. The simple model that will provide an estimate within 10 percent is often better than a complex one that will estimate within 1 percent but takes ten times longer to set up and run.

Table 5-1 shows the classes of prediction tools (see Sec. 5.1.2) that can be used, the model inputs (see Sec. 5.1.1), the data required, the limitations, accuracy, and time required to build the model.[1]

5.1.1. Variation Models

A variation model describes, in a quantitative form, how variation in individual parts and process will impact the value of the system KCs. The model can take one of several forms including performance, sensitivity, and assembly models. Sometimes these models are described as *transfer functions*.

---

[1]In some cases, there will just not be enough information. The product or manufacturing technology may be too new. The product development team may be unwilling or unable to quantify the risk because of the amount of uncertainty. In these cases the team should make rough estimates that bound the minimum and maximum probability.

Table 5-1. Model types, inputs, and limitations

Prediction Tool	Model	Input Data	Limitations	Accuracy	Time to Build Model
Previous designs	Performance of previous design	Previous designs	Based on engineering judgment	Fair	Low
Qualitative	Performance of previous design	Engineering judgment	Not based on numbers	Fair	Low
Extreme value analysis	Performance model	Process capability	Gives the worst case, but simple to analyze	Fair	Med
Tolerance validation	Tolerance model	Engineering tolerances Process capability	Dependent on correct allocation of tolerances and availability of accurate process capability data	Fair	Low
Root-sum-squared	Sensitivity model	Process capability	Can be time consuming to build sensitivity model	Good	Med–High
Monte Carlo	Performance model	Process capability	Time consuming to run the model	Excellent	High
Geometry-based variation simulation software	Assembly model	Process capability	Useful only for geometry	Excellent	High
Statistical correlation	Statistical model	Previous designs	Shows only correlations, not causality	Excellent	High
Prototype	Prototypes	N/A	Only as good as prototype	Good–excellent	High

Performance Model

A performance model predicts the actual value of the system KC based on the input part and process KCs values. The performance model is typically used in the Monte Carlo approach and tolerance stack-up analyses described below. A simple example is an equation where current is a function of voltage and resistance:

$$I = \frac{V}{R} \tag{5-1}$$

Suppose the target voltage is 10 V, resistance is 2 Ω, and current is 5 amps. If the voltage is high at 10.5 V, the resulting current is 5.25 amps—5 percent above target. This function allows for exact calculation of the final system KC value. However, in order to compute the distributions of the system KC, it is necessary to run many simulations using a Monte Carlo approach (see the description later in the chapter). A mathematical performance model may include complex equations that require numerical solutions, such as finite element analyses or performance simulators such as ray tracing.

Sensitivity Model

The sensitivity model measures how much the system KC will vary from its target value given a small deviation of the part KCs. This model is used in the root-sum-squared approach. In our example, to directly estimate without a Monte Carlo simulation, the variability in current, it is necessary to know how sensitive the current is to variability in voltage and resistance. For example, current will vary given a small change in the voltage or resistance and can be calculated by taking the partial differentials of the governing equations (also called a Taylor expansion).

$$\frac{\partial I}{\partial V} = \frac{1}{R} = 0.5 \text{ amps/V} \qquad \Delta I = 0.5 \Delta V$$

$$\frac{\partial I}{\partial R} = \frac{V}{R^2} = -2.5 \text{ amps/} \Omega \qquad \Delta I = -2.5 \Delta R$$

(5-2)

Equations 5-1 and 5-2 show that if the voltage varies by 0.5 V from target, the current will be 0.25 amps too high. These sensitivities are also used to compute the relative contributions to system variation described later in the chapter. Only the first-order sensitivities are shown here. A higher-order analysis using partial differentials provides the second-order and coupled sensitivities, resulting in a more accurate assessment. Design of Experiments on simulation models or physical prototypes can also be used to determine the sensitivities. Sensitivities can also be estimated using engineering expertise, tolerance stack-ups, or error budgets.

Assembly Model

Assembly models are geometry-based models that describe how two or more parts will assemble and how the final characteristics of the assembly are generated. It is possible to determine how small variations in the geometry of parts will impact the final dimensions of the assembled product without deriving the governing equations which can be difficult to generate by hand.

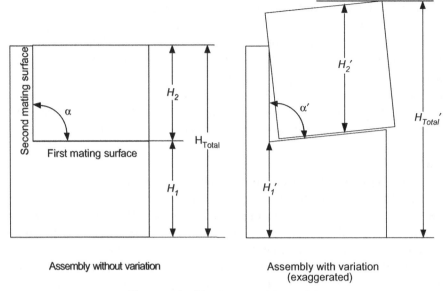

Assembly without variation   Assembly with variation
                                                  (exaggerated)

Figure 5-3. Simple assembly model.

Figure 5-3 shows a sample geometric assembly. The variability of the total height is the variable of interest and will be determined by variations in the angle α and in the heights H_1 and H_2. When modeling the impact of variation, it is necessary to describe both the geometry and how parts are assembled. For example, if the order of assembly in the example were changed making the vertical surface first and the horizontal second, a different total height would result. Typically a Monte Carlo simulation is used to predict the final output; however, some geometry variation tools automatically create linear models based on the equations governing the assembly and use root-sum-squared approaches.

Tolerance Model
In this model type, the IPT sets tolerances on the individual parts such that, if the part and process tolerances are achieved, then the system tolerances are met.

Performance of Previous Products
In some cases it is not possible to model the complex interactions between design parameters and product performance. However, it may be possible to use the history from previous designs to predict the performance of future designs. The performance can be analyzed quantitatively or estimated by using engineering judgment and experience.

Prototypes

In some cases it is not possible to quantitatively model the product because there is not enough information. This typically happens when new technology is being developed. The prototypes can be used to test both the performance and the sensitivity to variation.

5.1.2. Prediction Tools

Once the available models have been identified, a predictive tool to operate on the model must be determined. This section reviews several predictive tools, including:

- **Similarity to Previous Designs.** This approach evaluates the difference between previous designs and the proposed design to predict the defect rate of the new product.
- **Qualitative Assessment.** Engineering judgment is used to rank the defect rates when no quantitative model is available.
- **Tolerance Validation.** This method assesses the capability of achieving the individual tolerances given the current process capability.
- **Extreme Value Analysis.** This provides the worst-case analysis for variability. It is relatively simple but greatly overestimate the expected variation.
- **Root-Sum-Squared Analysis.** RSS analysis is a statistically based analysis, provides a relatively accurate assessment of the total variation, and requires little computational power.
- **Monte Carlo Simulation.** This numerical simulation is very flexible but can be time consuming to run.
- **Geometric-Based Variation Simulation Software.** This simulation uses the geometry model to predict the dimensional quality of assemblies.
- **Statistical Correlations.** The correlation approach can be used when no variation model is available but there is significant information about a large number of previous designs.
- **Prototypes.** Physical prototypes can be used to simulate product performance including tolerance stack ups.

Similarity to Previous Designs

In some cases, the team can use yields from previous designs to estimate future yields. The team should collect data from existing products and determine the differences between the previous design and the new design. Based on engineering analysis and/or judgment, teams can estimate the expected increase or decrease in the defect rates.

For products that have the same part types in different configurations (for example, different disposable medical products) or that have similar manufacturing processes (for example, hole drilling). The team can predict future defect rates by normalizing historical data by:

- Computing the total defects for each system KC for the existing product.
- Estimating the improvement from better manufacturing processes, more robust designs, and so on.
- Normalizing the number of defects for the new product. For example, if the previous design had 100 holes, and the new one has 50, this would halve the expected defect rate for hole drilling assuming the processes and target tolerances are similar.

Qualitative Assessment

If there is no previous design, and/or the system is impossible to model, then a qualitative assessment of defects in system KCs can be made. The team should assign a 1-to-10 measure (similar to an FMEA) based on the relative probability of a defect. Table 5-2 contains guidelines for assigning a qualitative value for the probability of exceeding the allowable tolerance. If a team feels uncomfortable picking a single value, it can assign a range. In addition, teams can use engineering estimates for C_{pk} (based on experience) to assign an estimated number. In most, cases, team members will know which processes are likely to have a C_{pk} of 1.33 or better and which have values significantly below 1.0. (See Chap. 2 for a definition of C_{pk}.)

Tolerance Validation

During product development, the IPT will allocate tolerances to the individual parts such that, if the tolerances are satisfied, the system KC should be acceptable.

One common misconception held by many teams is that they can evaluate the quality of the individual parts, and that, if the parts meet their targets, the

Table 5-2. Qualitative assessment of probability of a defect

Qualitative Assessment	Criteria	C_{pk}
1	Low probability; almost never happens	$C_{pk} > 1.33$
4	Rare, but a risk	$1 < C_{pk} < 1.33$
7	Likely to happen	$0.5 < C_{pk} < 1.0$
10	Almost certain to happen	$C_{pk} < 0.5$

final quality of the assembled product will be acceptable. Some companies set criteria that all parts must have a C_{pk} of greater than 1.33 and assume that the assembled product will also have a C_{pk} of greater than 1.33. However, two factors can make this assumption invalid. First, tolerances on the system may not be set to be consistent with the part tolerances. Second, the mean shifts of several parts in an assembly, even if standard deviations are small, can add up in such a way that the assembly C_{pk} is adversely affected. In cases where parts are machined, the mean shifts can be a significant contributor to this type of problem. Often operators will machine parts to the "maximum material condition." This is done to hedge against an error in machining. If an error occurs, the operator has the entire tolerance available to fix the problem. However, with the advent of highly accurate numerically controlled (NC) machining, the chance of needing the excess tolerance is relatively small.

Table 5-3 shows a simple stack-up example in which three parts contribute variation to a final assembly. All three parts fall within the acceptable limits given the standard deviation and mean shift. If all three parts had no mean shift, the final assembly would have a high C_{pk}; however, because the mean shifts all add up in the same direction, the resultant C_{pk} is significantly lower than the required 1.33.

Extreme Value Analysis

The extreme value approach calculates the worst-case scenario for the assembly. It uses the worst-case values for all KCs to determine the maximum deviation for the system KCs.

Figure 5-4 shows how the upper and lower limits for the entire assembly are determined based on the upper and lower limits of each part. If a performance model is available, it can be run with all worst-case extremes to calculate worst-case upper and lower limits on the system KCs. Because the probability of all contributors being simultaneously on a lower or upper limit is small, this type of analysis provides an overly conservative estimate of the expected variation.

Table 5-3. Effects of mean shift on C_{pk} of assemblies

Part	Nominal	Sigma (σ)	Mean Shift (b)	Tolerance	C_{pk}
1	9	0.2	0.2	±1	1.33
2	5	0.2	0.2	±1	1.33
3	1	0.2	0.2	±1	1.33
Assembly	15	0.35	0.6	±1.7	1.05

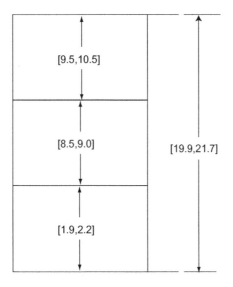

Worst case (1)
            9.5+8.5+1.9=19.9
Worst case (2)
            10.5+9.0+2.2=21.7

Figure 5-4. Extreme value analysis.

Root-Sum-Squared Analysis

The more statistically valid way to evaluate tolerances is using the root-sum-squared (RSS) approach. When combining two features that vary according to Gaussian distributions, the distribution of the combination also has a Gaussian distribution. The RSS approach requires the use of a sensitivity model.

The RSS approach is based on the sensitivities of a system KC and to the expected variation in the contributing part and process KCs. The sensitivities define how much a system KC deviates from the target value given a small deviation in a contributing KC. The example used earlier of the electric current is used again to demonstrate the RSS approach.

To estimate the variability in current, it is necessary to know how sensitive the current is to variability in voltage and resistance. Repeating equation 5-2, the sensitivities were calculated by taking the first-order partial derivatives (also called the Taylor expansion).

$$S_V = \frac{\partial I}{\partial V} = \frac{1}{R} = 0.5 \text{ amps/V}$$

$$S_R = \frac{\partial I}{\partial R} = -\frac{V}{R^2} = -2.5 \text{ amps/}\Omega$$

(5-3)

The system KC's standard deviation σ and mean shift b can be calculated from the standard deviations σ_i and mean shifts b_i of the contributing KCs and the sensitivities S_i using an RSS calculation.

$$\sigma^2 = \Sigma(S_i\sigma_i)^2, \quad (\sigma_I)^2 = (S_V\sigma_V)^2 + (S_R\sigma_R)^2$$
$$b = \Sigma S_i\, b_i, \quad b_I = S_V b_V + S_R b_R \tag{5-4}$$

The team should determine the capability of the manufacturing processes based on industry standards, machine data sheets, manufacturing process capability databases, and manufacturing expertise. Appendix B describes how to store, locate, and utilize process capability data from the factory for VRM during product development.

Table 5-4 shows the example of a simple circuit. The percent contribution will be calculated in the section below.

Once the system KC standard deviation and bias are calculated, it is possible to calculate the defect rate (Fig. 5-5) by calculating the percent of the distribution that falls inside the upper and lower limits (tolerance or margin).

If the allowable variation is ± 0.25 amps, the probability of exceeding the upper limit is close to zero and that of exceeding the lower limit is about 1 percent. The total yield would be 99 percent (defect rate of 10,022 defects per million). In addition, the team can calculate C_{pk} from the predicted variation.

The C_{pk} in the example would be 0.77, so the manufacturing process, with the given variation, does not satisfy the typically acceptable C_{pk} minimum of 1.33.

The RSS calculation is good for estimating the expected variability in system KCs assuming errors are independent and are relatively small. A downside is that the time to set up the analysis can be significant and the accuracy of the results is highly dependent on good information about sensitivities, manufacturing process capabilities, and tolerances. In addition, RSS calculation can only be used with process variations that follow a Gaussian distribution. In cases where there are multiple layers in the variation flow-

Table 5-4. Data for simple circuit

	V	R	I
Nominal	10 V	2 Ω	5 amps
Sensitivity (S)	0.5 amps/V	-2.5 amps/Ω	
Standard Deviation (σ)	0.1 V	0.01 Ω	0.056 amps
Bias (b)	0.01 V	0.05 Ω	-0.12 amps
Percent contribution	13%	87%	

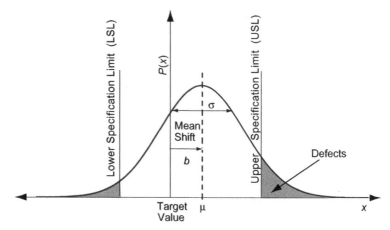

Figure 5-5. Gaussian distribution with a tolerance.

down, the RSS analysis will need to be completed for each layer working upwards until the system KC variation is predicted.

Monte Carlo Simulation

In cases where it is not possible to determine sensitivities or when the distributions are not Gaussian, the impact of variation may need to be simulated. The method most typically used is the Monte Carlo simulation. The inputs to the simulation are the process capabilities for the part KCs (Chap. 12). The statistical distributions of the part KCs are typically modeled as Gaussians for simplicity, but any type of distribution can be used. For each iteration, the simulation randomly assigns a value to each of the part KCs based on its statistical distribution, and the resulting system KC value is calculated from the performance model. After thousands of iterations, the result is a distribution of system KC values that is analyzed and compared to the system specification to generate an expected defect rate.

In Fig. 5-6, two parts are being assembled. The feature of interest is the gap between the two parts. The Monte Carlo simulation varies sizes of the input parts, models their assembly, and, after many iterations, computes the output variability.

This method has several benefits: It is easy to encode, can be used on noncontinuous, geometric, and complex models, and can combine a variety of distributions easily. In addition, a number of off-the-shelf software packages are available to automate the simulation and summarize the results. The major shortcoming of the Monte Carlo simulation is the time required to run the thousands of calculations required to get a statistically significant result.

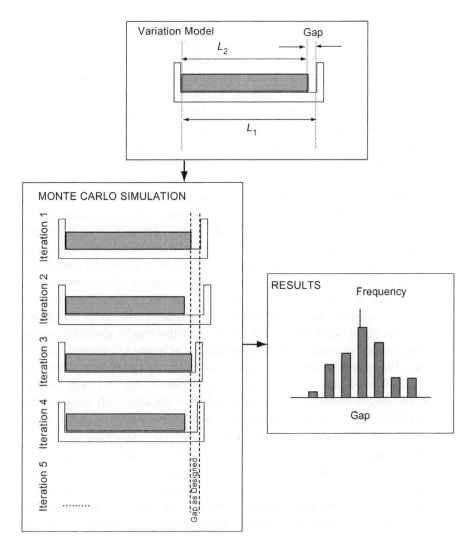

Figure 5-6. Monte Carlo simulation.

Geometry-Based Variation Simulation Software

In the sections on extreme value analysis, RSS, and Monte Carlo simulation, the team is guided to build an equation-based variation model. One specific case that regularly occurs in product development is the impact of geometric variation.

The equations governing final assembly dimensions are too complex to generate with reasonable effort. There are a number of tools that are specific geared to modeling variation in the geometry of assemblies and systems. Three examples of commercially available software are Vis-VSA, Dimensional Control Systems, and CE/TOL. All three tools use a computer-aided design

(CAD) model of an assembly as an input. The user defines variation in the dimensions as well as how parts are mated together and the order of assembly. Variation in each part is defined by specifying tolerances and a capability level (distribution, mean shift, and standard deviation). The software packages use a variety of techniques, including linearization and Monte Carlo simulation, to predict the resultant variation in the assembly. The tools typically provide analysis of the relative contributions of variations in the dimensions of parts to system defects. This type of simulation is very effective for simple assemblies but are limited in the following cases.

- **Compliance Parts.** In many applications, parts cannot be assumed to be rigid. For example, the skin panels on an aircraft wing have a great deal of compliance, and their shapes can be altered with minimal force. When the wing is assembled, the compliance is used to ensure that the skins are flush with the ribs and spars. If the wing was modeled as a series of rigid parts, the predicted quality would be very poor. Modeling compliant parts is made even more difficult when there is a maximum allowable force allowed to assemble parts. For example, the hydraulic tubes in an aircraft are very compliant but there is a limit to the force that can be applied to mount them into the structure. Several approaches can be used to address compliance. First, the compliant part can be modeled as two parts with a bend line modeled as a hinge. Second, simple finite element analysis can be used to model the compliance. The first approach does not accurately represent the assembly, but the second can be very time consuming if a Monte Carlo simulation is being used and the FEA needs to be run multiple times.

- **Overconstrained Parts.** In some cases it is necessary to design overconstrained assemblies—that is, there are more mating constraints than there are degrees of freedom. Bolt hole patterns are a classic example. In Fig. 5-7, the bolt hole pattern on the left is properly constrained. The lower pin constrains the assembly in the x and y directions and the pin

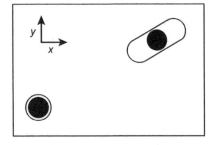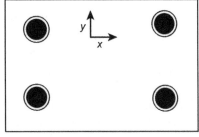

Properly constrained   Overconstrained

Figure 5-7. Correctly constrained versus overconstrained assembly.

in the slot constrains the part rotationally. As long as the pins are within a certain (relatively large) distance of each other, the two parts will assemble. However, bolt hole patterns like that on the right are often specified because of the need to sufficiently clamp the two parts together. In this case, the locations of bolts and holes must be tightly controlled, otherwise the two parts will not assemble. One of two variation modeling approaches can be used here. The first models the mate as being constrained by two of the bolts and holes along their centerlines. The software then checks for interference in the two other bolts. This approach can lead to a higher predicted defect rate than will occur in reality. The second approach does a local optimization to find the best orientation of the two parts while mating the holes. This approach can be computationally intensive.

- **In-Process Adjustments.** Shimming, trimming, tuning, and selective assembly are all examples of in-process adjustments. Building models of these assembly techniques can be challenging when using the traditional geometric variation tools.

While these tools are powerful at simulating variation in assemblies, they are limited to modeling products where detailed geometric models are available and are difficult to use early in product development. Another common pitfall is that teams tend to model everything in the product rather than focusing on one or two critical areas. Overly complex models become unwieldy, and understanding the results and then taking action can be difficult. The key in modeling variation in geometry, as with any model, is to pick only the critical areas where questions need to be answered.

In addition, the results are only as good as inputs to the models. Geometric variation modeling tools can be used to evaluate the tolerancing of a product; however, the results need to be integrated with good information about process capability in order to predict the actual system performance.

Statistical Correlations
In some cases, it will not be possible to create either a sensitivity or performance model to predict the yield directly. Prediction tools based on statistical correlations can be used in this situation. It is known that certain design parameters are more likely than others to impact robustness. For example, for printed circuit boards, the yield and reliability per unit area of the PCB are functions of the individual solder joints, the reliability of chips and other components, and any bridging (shorts across leads) that may occur (Nagler, 1996). Using data on previous designs produced using similar processes, it is possible to correlate the start-up and ongoing yields to the design characteristics. For example, several factors have significant impact on the yield, including:

- **Number of Sides.** On PCBs with components on both sides, the chance of damaging a connection or component is increased.

- **Number of Hand-Loaded Parts.** If parts cannot be placed using automated equipment and have to be hand located, the yield drops significantly.
- **Line Pitch.** *Pitch* refers to the space between leads on components and between lines on the board. The finer the pitch, the more likely it is that a short or a bridge will develop.
- **Part Density.** Higher densities of parts per unit area lead to lower yields.
- **Process.** Most companies have multiple process lines that can run different configurations and different products. Each line has a different process capability.

By plugging in the design parameters for a new PCB, it is possible to determine approximate yields. The uncertainty in the predicted yield can be sizable because of other, nonquantifiable differences between boards. However, the accuracy of the yield predictions is not critical, this kind of system can be used to identify the changes that have the biggest impacts on yield. It is important to remember that the model generated using statistical correlations only provides data on correlative behavior, not causality.

The method for creating the statistical correlation between the design parameters and the start-up and long-term yields is as follows:

- Identify the metrics you are designing for (yield, throughput, etc.).
- Identify what design/process characteristics are (1) likely to have an impact on the metrics and (2) able to be changed.
- Identify what data is available from previous designs. If data is not available, then put in place a plan for collecting historical or present process data.
- Run a statistical analysis on the data to determine first-order and second-order correlations (Box et al., 1978).
- Verify the correlations on the new designs.
- Put a plan in place to use the analysis early in product development to identify what products are likely to have issues in production.

Prototypes
As was pointed out earlier, it is often difficult to accurately assess the risk associated with some KCs. The product may include new technology whose behavior and variation are not understood. The manufacturing processes used to make the product may be new and the variation introduced may not be quantified. If the uncertainty and cost of exceeding the allowable variation are too high, the team may need to invest in prototypes.

A number of different prototyping strategies can be implemented (Ulrich and Eppinger, 1995):

- **Concept prototypes.** Concept prototypes evaluate the product's function but are not built using the manufacturing techniques used in full production. Concept prototypes can indicate where variation is likely to impact the final performance by allowing the team to better understand how the product functions; however, because the final production techniques are not employed, it is difficult to determine whether there will be excessive variation during production.

- **Manufacturing Prototypes.** These prototypes typically are small production runs on the actual or similar production equipment. They are used to evaluate whether the current equipment can manufacture the product with the appropriate quality. The data collected on manufacturing prototypes is typically representative of the short-term process capability, which may be better than the long-term capability.

5.2. ESTIMATING THE CONTRIBUTIONS OF PART AND PROCESS KCs

Once the relative defect rate of each system KC is determined, it is necessary to determine how to fix the problems. Reducing the defect rate can occur via one of two options—reducing the sensitivity of the system KC to variation or reducing sources of variation. Either way, it is necessary to determine which part and process KCs make the largest contributions to the system KC defect rates. If a system KC's variation is too high, it is necessary for the team to look at reducing variation in the part and process KCs that contribute to the system KC. The team is more likely to have success reducing variation in a system KC by working on a part or manufacturing process that contributes 70 percent of variation than by working on one that contributes only 10 percent.

There are two aspects to a variation contribution: expected variation and sensitivity. The expected variations are mean shifts and standard deviations of the contributing part and process KCs.

The sensitivity measures how much a system KC varies given a small variation in a contributing KC. These sensitivities are the same as those calculated in Sec. 5.1.1 and used for the RSS calculation. Figure 5-8 shows two examples of different sensitivities. In the stack-up shown on the left, a single unit change in h_a will result in a single unit change in h. However, in the example shown on the right, a small change in the angle α can result in a large change in h.

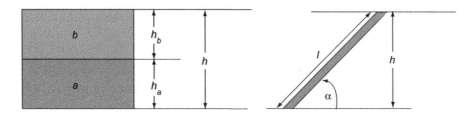

Sensitivity to a height change:
sensitivity = 1

Sensitivity to an angle change:
sensitivity = $L \cos \alpha$

Figure 5-8. Sensitivity as a function of geometry and assembly.

The example of the optical collimator from Chap. 2 is used to demonstrate the usefulness of computing the relative variation contributions. The next section provides information on how to calculate the contributions. Figure 5.9 illustrates the variation flowdown from the system KC (optical losses) to the controllable part and process KCs. At each layer, the contribution of each KC to its parent KC is determined. The end leaves of the flowdown branches are those parts or manufacturing processes that can be controlled either by the supplier or by in-house production. Figure 5-9 shows that 50 percent of the variability in the objective lens loss occurs due to the transmission, and 50 percent is due to focusing losses. Forty percent of the variation in the optical transmission is caused by contamination.

The total contribution of each controllable KC is found by multiplying the percent contributions of each KC on the branch. For example, the in house sources of contamination contributes 4 percent to objective lens losses (50 percent transmission losses × 40 percent contamination × 20 percent in house contamination). Figure 5-10 shows the relative contribution of each

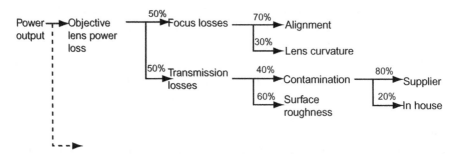

Figure 5-9. Contribution of part KCs to overall variation in power loss in a collimator.

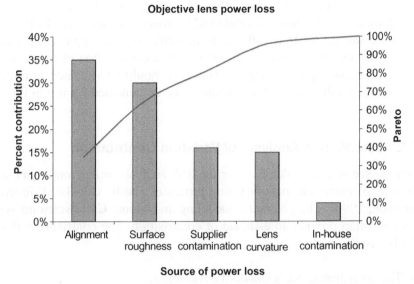

Figure 5-10. Relative contribution of part KCs to power loss in a collimator.

part and process KC. Alignments are the largest contributor at 35 percent and is in the factory's control. However, the contamination introduced by the supplier (18%) may not be as easily controllable.

One of two methods can be used to determine the percent contributions: qualitative or quantitative.

5.2.1. Qualitative Analysis of Variation Contribution

In cases where quantitative models are not available, the IPT may use qualitative estimates of contribution. Based on previous experience, the team should estimate the relative contributions of part and process KC variations to the system KC variations. Minimal contributors should be in the range of 1 to 10 percent. Contributors that are major drivers should be in the range of 70 to 90 percent. The team should specify the contribution of each KC such that the total sums to 100 percent.

When deciding how to assign the relative contributions, the team should ask the following two questions:

1. How sensitive is the KC to the contributing KC? If there is a low sensitivity, then the overall contribution is likely to be small.
2. How much variation is introduced by the contributing KC? If there is significant variation in the contributing KC, then the contribution is likely to be large.

An example where it is often difficult to predict the relative contribution through models is the noise vibration and harshness measure used by the automotive industry. NVH is difficult to quantify, and the engineering team may need to use experience to rank the relative contribution of the various sources. The assumptions about contribution should be validated in prototype testing and the lessons learned should be documented for use in future programs.

5.2.2. Quantitative Analysis of Variation Contribution

Using the sensitivities calculated for the RSS analysis earlier, variation contributions of parts and manufacturing processes can be calculated directly (Thornton, 1999). The geometric modeling and Monte Carlo software systems will often provide an estimate of contribution. The contribution P_i of KC i is based on:

- The contributing KC's standard deviation σ_i
- The contributing KC's bias b_i
- The parent KC's sensitivity to the contributing KC S_i

The relative contribution of KC i is calculated by:

$$P_i = \frac{(S_i\sigma_i)^2 + (S_ib_i)^2}{\Sigma(S_j\sigma_j)^2 + \Sigma(S_jb_j)^2} \tag{5-5}$$

where j represents all of the contributors to the system KC.

The results for the example from the simple circuit analysis are that variation in the resistance will contribute about 87 percent and the voltage 13 percent. It is important to include both the bias and the standard deviation in the calculation because both can have a significant impact on the overall variation of the system KCs.

5.3. MEASURING THE FREQUENCY OF DEFECTS

For existing products it is not usually necessary to predict the frequency of defects because the data should exist. Sources of defect data include:

- **Financial Reports.** These can include gross numbers on parts and materials received and good product out of the factory. Sample financial data is shown in Table 5-5. These numbers provide the gross measures of product loss but do not provide details about where or why parts are not used in shipped product.

Table 5-5. Sample financial data

	Parts In	Parts Out
Needles	5000	4500
Tubing	5000	4300
Cannula	5000	4200
Assembly	4200	4000

- **Scrap/Rework Reports.** Individual assembly and inspection stations may track which parts, materials, assemblies, systems, and products are scrapped, the reason for failure, and the action taken. However, the data may be incomplete; and it may be necessary to go to the production floor to fill in any gaps. Data about scrap is often maintained in many locations, and all the data must be integrated to properly assess the defect rates.
- **QC Plans and Reports.** Quality control data is maintained by a number of organizations including production, quality control, regulatory oversight, finance, and research and development (R&D). An example of incoming inspection data is shown in Table 5-6.
- **Customer Complaint and Warranty Data.** This data is useful for measuring product defect rates and, if specifics on failure modes are recorded, for analyzing causes of defects.

While these data sources provide a good starting point, they do not present a complete picture of all impacts of part and process variations on system variations. In order to thoroughly identify all sources of variation, it is beneficial to create a map of the production processes and analyze, for each step, whether variation drives cycle time, scrap, rework, throughput, labor time, inspection, or testing requirements. There are many books on process mapping and business improvements (Galloway, 1994; Womack and Jones, 2002). Figure 5-11 shows a simple example of a process map with the standard process symbols. When mapping a product's manufacturing processes, you should include:

- **All Processes and Work Done on the Product.** Every step including inventory transport, rework, and operations should be included.

Table 5-6. Sample incoming inspection data

Needles	
Total good	96%
Scrapped due to PM	4%

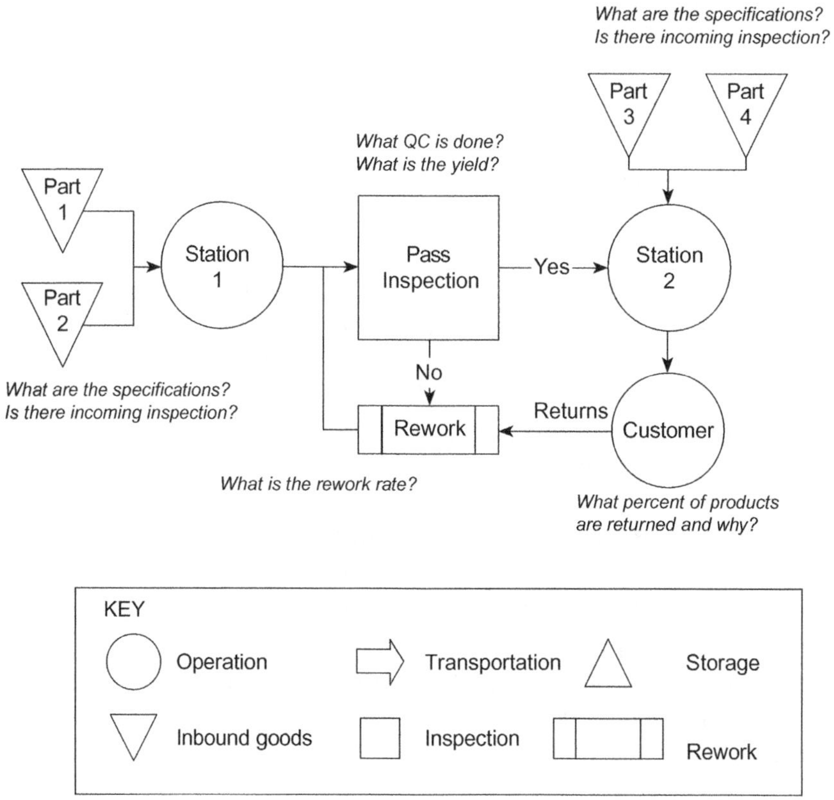

Figure 5-11. Sample process flow with sources of defects identified.

- **All Incoming Parts and Materials.** Incoming materials may be inspected and/or measured. The defect rate for incoming parts may be already tracked.
- **Quality Control.** These are the points where scrap and rework are generated. The scrap and rework rates should be identified.
- **Rework Stations.** These are the stations where the rework occurs. There may be quality checks at these stations as well.
- **Customer Feedback.** Customer feedback may include warranty returns. Returns may or may not be reworked.

The data for each source of defects will be combined with the cost data described in the next chapter.

Financial reports often will show significantly more product loss than the manufacturing process records will show. Tables 5-5 and 5-6 show an example of such a discrepancy. The financial data shows that during incoming inspection, 4500 of 5000 parts ordered were passed to the assembly operation (a 10 percent loss in the incoming inspection area); however, data collected in the incoming inspection area only shows reasons for a portion of the scrap (i.e., 4 percent). Such discrepancies can be due to:

- **Incomplete Scrap Data.** Not all inspection and QC stations record data. All causes for scrap and loss should be identified and tracked.
- **Ignoring Setup Costs.** Scrap often occurs during the setup of a production process or during a shift change. For example, one company was scrapping a significant amount of plastic sheeting used to build a medical product. The financial reports showed a high scrap rate, but the scrap rate shown by the manufacturing process records was an order of magnitude lower. It was found that the first section of material on the sheeting rolls had to be scrapped due to damage incurred during shipping. The station never recorded this waste, even though it represented a significant portion of the scrap cost.
- **Ignoring Tests and Sampled Product.** Organizations will often lose products to destructive testing sampling. Typically, the product pulled for testing is fully assembled and, as a result, expensive to destroy. Evaluating whether the number of sampled units can be reduced can result in a reduction of the total manufacturing cost.
- **Ignoring Product that Falls on the Floor.** Especially in the case of medical products, any product that falls on the floor is often just thrown away and the value of the material is never recorded. It is important to determine losses due to issues such as handling.

When there are significant discrepancies between the data sets, it may be necessary to add additional data collection to the factory floor to refine the information being collected. Additional data should be collected (1) where there is a large rate of defects and (2) where the data has very little resolution.

When measuring the defect rates, it is necessary to pick a standard population size over which to average the defect rates. It is possible to average over a batch week, or year. When determining the statistically valid period of time or population to use, the team should determine how the data is going to be used, how often it will be updated, and how large the data set is.

If customer complaint or return data is not available, the internal defect rates are often a good proxy for defects that escape to the field. This occurs for two reasons. First, quality control will let a certain number of bad products through. To reduce the certainty of bad products passing through, the organization will either need to implement multiple layers of quality control

Table 5-7. Parts near tolerance limits

Defect Rate	Remaining Parts Within 10% of Tolerance Limits
0.09%	0.68%
0.13%	1.37%
1.24%	3.35%
3.21%	5.62%

and/or risk rejecting a large number of good parts. Second, if products exceeding defined limits are rejected or reworked, there will be a number of products that fall just inside the acceptable limits. These products are more likely to cause customer complaints and/or returns. Table 5-7 shows for a set of defect rates how many of the remaining parts fall within 10 percent of the tolerance limits. As is shown in the table, the larger the defect rate, the larger the number of products that are very close to unacceptable.

5.4. MEASURING THE CONTRIBUTIONS OF PART AND PROCESS KCs

When looking at products in production, the IPT needs to investigate the major contributors to the defect rates in system KCs. Depending on the completeness of the quality system, the data being collected already, and the ability to model the product, determining the contributions may be easy or may require significant work.

The team should focus its efforts on understanding the major contributors to variation on those system KCs that have both a high defect rate and a high cost of defects. By examining data and running experiments it is possible to understand (1) where system KCs are highly sensitive to variation in the part or manufacturing process KCs and (2) where there is significant variation in the part or manufacturing process KCs. There is a huge literature base dealing with design of experiments, capability studies, and Six Sigma that can provide details on how to do this type of analysis.

- **Capability Studies.** The individual part and/or process KCs should be measured over a period of time to understand the mean and standard deviation of the characteristic. The team can draw on existing SPC data, previous capability studies, or other scrap and rework data as a starting point.
- **Design of Experiments.** This tool combines various settings and/or parameters to understand experimentally how sensitive the system KC is to the part and process KCs.

- **Variation Modeling.** If using design of experiments is too time consuming, the team can use anything from back-of-the-envelope calculations to complex computer simulations to understand how system KCs are influenced by the part and process KCs.

5.5. SUMMARY

The result of the assessment of the frequency of product (or system) defects and the contributions of part and process KC variations allows the team to focus on areas of highest defect rates and on what parts and manufacturing processes contribute the most to those defects.

Assessment in the early stages of product development should use back-of-the-envelope estimates and graduate to increasingly more detailed methods as the product's design matures. One typical mistake organizations make is using a too detailed analysis on too many of the product's systems, assemblies, and parts. One of the skills of assessment is to know what *not* to include in the analysis to get the best information at the most reasonable cost. The goal of assessment is not to predict to the tenth of a percent what the yield will be but rather to identify what areas have the higher defect rates. Teams should use the 80-20 rule in assessing the product: Try to get 80 percent of the accuracy, with 20 percent of the effort.

Assessment during production should rely on existing data on defect rates, scrap, and rework. The team may need to augment the existing data with additional measurements and data collection. One typical mistake teams make is to try to detail too much of the defect rates and their causes. The team should do a broad assessment first to determine where the major drivers are and should only try to detail those areas where the cost is the greatest.

For both cases, the IPT should remember that the assessment is not the end goal or product. It is the action taken in response to the information generated that creates the value.

6

ASSESSMENT OF COST AND RISK

The previous chapter discussed how to assess the frequency of defects. The chapter is divided into sections on cost and risk analysis in product development and cost analysis in production. Everyone should review the section on cost sources in the production section to understand how broad an impact variation can have. Figure 6-1 shows the role of this step in the overall identification, assessment, and mitigation procedures.

The goal of the previous chapter was to enable IPTs to identify which system KCs have the highest defect rates and what the major contributors are. However, this information is not enough to prioritize which areas of a design require improvement or changes. For example, a product may have two system KCs, one with a defect rate of 15 percent and the other with a defect rate of 1 percent. At first glance, the team would be likely to focus on the first system KC. However, if costs are taken into consideration and the first KC costs $10 to fix while the second costs $500, the second problem is clearly a higher priority. In order to properly assess the product, variation costs must be assessed.

In assessment during product development, the ranking is based on risk, or equivalently the expected cost of variation. The IPT needs to predict the cost and frequency of each defect to compute the expected cost.

In assessment during production, the ranking is based on the measured total cost of variation. The IPT can collect real data about costs of scrap, lost capacity, and so on, and relate that data back to variation in the manufacturing processes. This is not to say that calculating variation costs is trivial; collecting and understanding all cost drivers can be time consuming and challenging. In most cases, the data is incomplete, is owned by multiple people, and exists in many different forms and structures.

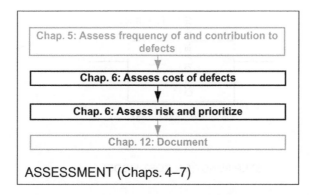

During product development

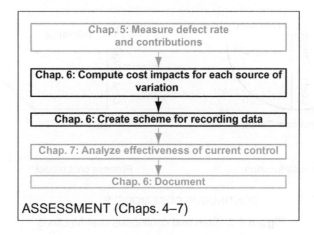

During production

Figure 6-1. Role of cost and risk during the assessment procedure.

6.1. COST AND RISK ASSESSMENT DURING PRODUCT DEVELOPMENT

In product development, there are two steps to calculating the risk (or expected cost): predict the expected defect rate in system KCs (Chap. 5) and predict the cost of a defect. Chapter 5 described the first. The three methods of computing the cost of a defect are qualitative methods, step cost functions, and continuous cost functions (Fig. 6-2). Step cost functions assume that the cost is only incurred when variation falls outside of the allowable tolerance (i.e., go/no go). Continuous cost functions assume that if the KC's value is exactly on target there is no cost, but that the cost increases as the deviation from target increases (e.g., the Taguchi loss function or process cost model).

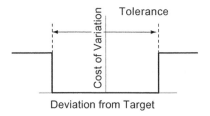

STEP FUNCTION COST MODEL

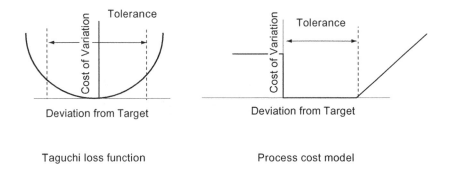

Taguchi loss function   Process cost model

CONTINUOUS COST MODELS

Figure 6-2. Step and continuous cost functions.

It is important to remember that as you compute variation cost; you should be identifying the cost of low system KC quality, not of the part and process KCs. Variation cost estimates are used for two purposes:

1. **Calculating the Risk Associated with a System KC.** The risk measures are used to rank and bucket system KCs to identify which ones need further attention from the IPT. The cost estimate only needs to be good enough to ensure that the system KC appears in the right bucket.

2. **Selecting between Mitigation Strategies.** When selecting a mitigation strategy, it is necessary to understand the cost benefit of any change. For system KCs for which several mitigation strategies are being evaluated, it may be necessary to develop more detailed cost models.

Once the cost and frequency of defects is determined, the team should compute the risks or expected costs (described later) and rank-order system KCs to identify which are of low, medium, high, or uncertain risk. Figure 6-3 is a repeat of Fig. 4-1.

As was pointed out in the previous chapter, in some cases there will just not be enough information to assess risk. The product or manufacturing technology may be too new. The product development team may be unwilling or unable to quantify the risk because of the amount of uncertainty. Based on the minimum and maximum possible probability of defects along with a bounded estimate of cost, the team should determine if the system KC falls into the low-, medium-, or high-risk bucket. If the range of possible risk is too great, the system KC should be categorized in the uncertain bucket.

6.1.1. Qualitative Assessments

Section 6.2 reviews models that can be used to quantify variation costs. However, developing detailed cost models is often not possible in product development—there just isn't enough information or time. The team can use qualitative assessments to compute risk when quantitative models are not available. By putting a 1-to-10 measure on defect rate (described in Chap. 5) and a 1-to-10 measure on cost, the risks of system KCs can be ranked based on engineering judgment.

$$\text{Risk} = \text{Defect rate } (1–10) * \text{Cost } (1–10) \qquad (6\text{-}1)$$

Table 6-1 provides guidelines for assessing the relative cost of exceeding the allowable tolerance on a system KC. The team should first determine the effect of a defect (e.g., scrap, rework, or customer dissatisfaction). Based on the response, the associated column can be used. For example, if the entire product is scrapped when a defect is found, a ranking of 10 should be given. The absolute number is not important but the relative ranking of the system KCs is.

Figure 6-3. System KCs are bucketed according to relative risk.

Table 6-1. Qualitative assessment of defect cost

Cost Measure	Scrap	Rework	Customer Satisfaction
1	Low unit cost	Easy to fix	Not critical
5	Medium unit cost	Can be reworked/repaired, but expensive	Important to avoid
10	High unit cost	Expensive to repair	Critical to avoid

6.1.2. Step Cost Functions

The step cost function (also called a go/no go function) assumes that a cost is only generated when the performance target falls outside the acceptable range. For example, a medical product is scrapped if it fails the leak test, or a car door is reworked if the gap is unacceptable.

The risk C_{exp} is a function of the defect rate p and the cost of a defect C_D.

$$C_{exp} = pC_D \qquad (6\text{-}2)$$

In this case the expected cost or risk can be computed by simply multiplying the defect rate (the percentage of products falling outside the limits) by the cost of exceeding the limits (i.e., cost of scrap, rework, or customer returns).

6.1.3. Continuous Cost Functions

The calculation of expected cost using continuous cost functions should only be done for a small set of system KCs for which detailed cost models are needed. In most cases, the qualitative or step function models will be sufficient to (1) rank system KCs and (2) evaluate the benefits of different mitigation strategies. The following section should be read by those who are interested in learning how to do more accurate cost assessment. There are two types of continuous models that can be used to assess cost: the Taguchi loss function and manufacturing process models. For continuous cost functions, the integral of the cost function $C(x)$ times the probability density function $P(x)$ is used to compute the expected cost. Most of the time, the probability density function is assumed to be Gaussian.

$$C_{exp} = \int_{-\infty}^{\infty} C(x)P(x) \qquad (6\text{-}3)$$

Taguchi Loss Function
The Taguchi loss function states that the only time zero cost of variation is incurred is when the product is exactly on target. As the product deviates

from the target value, the cost increases. In this case the cost of deviating from the target by Δx, L, is a function of the cost at the upper and lower tolerance limits A, and the difference between the upper limit and the target value Δ. The average loss $\bar{L}$, the expected cost, is a function of the bias b and the standard deviation σ.

$$L = \frac{A}{\Delta^2}\,(\Delta x)^2$$

$$\bar{L} = C_{\exp} = \frac{A}{\Delta^2}\,(\sigma^2 + b^2)$$

(6-4)

While the Taguchi loss function is good for explaining the concept of driving to on-target values and the importance of variation, it does not accurately reflect the actual costs. This approach tends to significantly overestimate variation costs but is good conceptually when discussing losses incurred by variation.

Manufacturing Process Cost Model

In some cases it may be possible to construct the cost as a function of deviation from target based on the actual response to an out-of-spec part (Thornton, 2001a). The example in Figure 6-4 is used to illustrate the method. Two parts are located relative to each other using either locating features or a fixture. The gap g between the two parts is ideally zero, because as the gap increases, unit costs increase (i.e., manufacturing time may increase or warranty returns may increase). However, an impact condition—where the gap is less than zero—will necessitate scrap or rework.

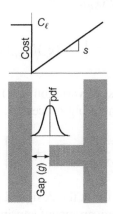

Figure 6-4. Manufacturing process cost model of a designed gap (Thornton, 2001a). Reprinted with permission of ASME.

This scenario appears frequently in a variety of industries. In aerospace, combinations of designed-in gaps and shims are used to mate parts and avoid an impact condition. When an impact condition occurs, parts may need to be reworked, trimmed, or, in the worst case, scrapped. In automotive components, one key quality characteristic of many automotive drive train components is durability. Durability is often a function of how closely parts fit together. However, if the gap is too small, parts may bind and not function. In the shipbuilding industry, welding is used to fill gaps between structural elements. Larger gaps take longer to fill and increase both the labor content and the cycle time. In some specialized applications, additional filler material also adds cost. However, if there is an impact condition between components, parts must be separated and trimmed, resulting in large rework costs. All cases are characterized by an unsymmetric cost/loss function. The larger the gap, the less likely an impact condition will occur; however, the smaller the gap, the lower the production costs.

The following shows the derivation used to calculate the expected cost of a gap assuming the gap varies with a Gaussian distribution with zero mean shift and standard deviation of σ. This derivation is given for those interested, but this type of analysis will typically not be done on a regular basis in the assessment phase of variation risk management.

If a gap of size g_p is produced, the cost of variation for that product $C(g_p)$ will be defined by the equation:

$$C(g_p) = \begin{cases} C_l & \text{if} \quad g_p < 0 \\ sg_p & \text{if} \quad g_p > 0 \end{cases} \qquad (6\text{-}5)$$

This cost function is shown Fig. 6-4. The cost if the gap is less than zero is C_l. If the gap is greater than zero the cost increases linearly with a slope s. The expected cost is found by integrating the cost function $C(g_p)$ with the probability density function $P(g_p)$.

$$C_{\exp} = \int_{-\infty}^{\infty} C(g_p) P(g_p) \, dg_p \qquad (6\text{-}6)$$

The resultant expected cost of variation is a function of the designed gap g and variation in the manufacturing process σ.

$$C_{\exp} = \frac{C_l}{2}\left[\text{erf}\left(\frac{-g}{\sigma\sqrt{2}}\right) + 1\right] + s\,\frac{g}{2}\left[1 + \text{erf}\left(\frac{g}{\sigma\sqrt{2}}\right)\right] \qquad (6\text{-}7)$$

This function can be used to set the designed gap g and tolerance to minimize the expected cost for a given manufacturing capability.

6.2. TOTAL COST OF VARIATION ASSESSMENT DURING PRODUCTION

For existing products, cost assessment is typically done assuming that costs are only generated when the KC exceeds the allowable variation (i.e., the step function or go/no go cost function). Cost assessment in production aggregates all costs throughout the organization associated with excessive variation in the KC. There are four steps:

1. Determine the cost sources and quantify them.
2. Develop a scheme for collecting the cost data and allocating costs to the individual KCs.
3. Aggregate costs to identify high-cost system KCs.
4. Identify key part and process KC contributors.

This data is used to rank the KCs as well as to feed the assessment of the current quality control plan (Chap. 7).

6.2.1. Cost Sources

The following sections outline the various costs that can be incurred as a result of variation. When calculating the total cost of variation, it is necessary to be consistent in the units used. One way of tracking the expected cost is to determine variation costs per good unit produced. However, when analyzing the total cost of variation including field failures, it may be necessary to look at returns on a per-unit time basis (e.g., per week) because the rate of usage in the field may be different than the production rate in the factory. As the cost is tracked, the team should make the denominators consistent to ensure that the results are valid.

Typically, when cost of variation is discussed, an organization will only look at scrap and rework. However, variation can have many hidden costs. Variation increases the overhead required to build a single good unit and decreases operational efficiency by increasing:

- **Scrap.** Throwing out a part, assembly, system, or product increases material, manufacturing processing, and labor costs.
- **Rework.** Fixing a defective part or system increases costs of labor, material, setup, and expediting.
- **Customer Dissatisfaction.** A defective product in the customer's hands incurs warranty costs and can lead to lost sales and legal expenses.
- **Inventory.** If there is scrap or rework, the factory must carry extra inventory to ensure steady flow in production and order fulfillment.

- **Lost Capacity.** If time is spent reworking or rebuilding product or in maintenance to reduce scrap rates, the factory's net capacity is decreased. Sudden degradation in quality has the potential for both starving upstream and blocking downstream stations. In addition, any scrap reduces the effective capacity of a machine.
- **QC Costs.** Organizations experiencing significant quality issues typically try to reduce defects by implementing quality control. Excess variation can increase the labor content (to conduct the inspection and measurement), overhead (engineering oversight), and capital expenditures (extra testing and measurement equipment), can decrease throughput, and can create bottlenecks.
- **Cycle Time.** The time to build a product from start to finish will increase if variation results in rework or additional touch labor.

The process map described in Chap. 5 (Fig. 5-11) can provide a starting point for identifying where there is excess cost (Fig. 6-5). When mapping a product's manufacturing processes, you should include:

- **All Processes and Work Done on the Product.** Excess labor caused by variation should be identified—for example, if a part takes a while to fit because it is too big and needs to be hand trimmed.
- **All Incoming Parts and Materials.** Incoming materials can be a major source of variation.
- **Inspection and Quality Control Steps.** These are points where scrap and rework are generated. Scrap and rework can add significant labor and increase the cycle time.
- **Rework Stations.** Often entire areas of a factory are dedicated to reworking product that did not pass inspection.
- **Inventory of Incoming, In-Process, and Finished Goods.** Often inventory is held to prevent disruptions to the factory caused by poor quality at various stages of production. The finance organization can calculate the cost of capital tied up in excess inventory.
- **Customer Complaints and Warranty Data.** When and why do customers return product or complain? (Chapter 9)

The next sections describe how to compute the cost of variation for a number of categories.

Cost of Customer Dissatisfaction

A number of products are inevitably returned to the manufacturer because of some real or imagined defect. These returns and/or customer dissatisfaction

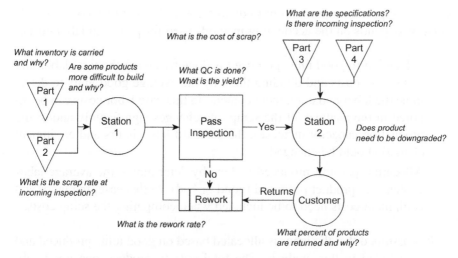

Figure 6-5. Process map used to identify sources of cost.

can be expensive. Product recalls often can incur not only repair and recall costs but also significant legal costs.

Not all defects will have the same cost associated with them. For example, discovering a defect in a medical product while it is being used on a patient is more expensive than discovering the defect when the sterile packaging is opened. To calculate the total cost, the team should assign a relative cost to each failure type C_{ci} based on the time, location, and type of failure, and multiply it by probability of that defect occurring in the population of shipped product p_{ci}. The expected cost per unit released to the fields is:

$$\Sigma p_{ci}C_{ci} \tag{6-8}$$

In production, it is often not possible to determine the usage rate in the field and hence the probability of finding a defect. In this case, teams may choose to track the total number of complaints over a period of time n_{ci}:

$$\Sigma n_{ci}C_{ci} \tag{6-9}$$

This number can be adjusted for seasonal trends.

Cost of Scrap
If a product, part, or assembly is scrapped, destructively tested, or lost, then the value at the point of scrap of should be calculated. When calculating the

cost of scrap, it is important not to double count the cost of labor. This calculation depends on the accounting methods used for parts and labor costs.

- Allocation to good units produced is done by taking the total labor and equipment costs and dividing them by the average good units to determine the labor and equipment content. In this case labor costs can be ignored in the analysis. If the scrap rate changes significantly and/or the production capacity increases, the standard labor hours and equipment charges should be changed.
- Allocation per unit produced is done by determining the average labor content per product produced (good or bad). In this case, the labor and equipment costs need to be included when computing the scrap costs.

It is assumed that the labor is allocated based on good units produced and is not included in this analysis. The total cost to produce one good subassembly (i.e., the adjusted cost) is a function of the cost of materials in the subassembly C_m and the yield rate y:

$$\frac{1}{y} C_m \qquad (6\text{-}10)$$

The cost of scrap for each good part produced is calculated by:

$$\frac{1}{y} C_m - C_m \qquad (6\text{-}11)$$

Risks of Standard Scrap

To account for scrap, organizations typically build in a standard scrap rate that allocates the cost of the rejected parts to the good parts produced. However, standard scrap rarely reflects actual scrap costs. Often, scrap costs for each part or subassembly are calculated at the end of the year and allocated as overhead. Often the scrap rates are then adjusted downward to meet product cost targets; however, these adjusted numbers are not communicated to the production groups as scrap reduction goals. The organization continues to pay for scrap that often shows up as general budget overruns. To effectively track and reduce scrap, target scrap rates and the way those targets are being met should be reported on a regular basis and managed.

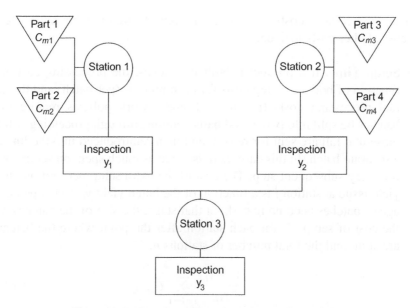

Figure 6-6. Process with multiple inspection points.

When computing the cost of scrap per unit produced, it is very easy to underestimate the cost. For each station, it necessary to use the adjusted cost of goods, not the cost of raw materials. For example, say one station has a scrap rate of 50 percent and the next station also has a scrap rate of 50 percent. In this case, to produce one unit, you need to start with four units, scrapping 2 at station 1 and one at station two. Figure 6-6 shows an example of a simple assembly process where two parts are assembled, inspected, assembled with two other parts, and then sent to final inspection. The scrap cost per good unit produced is equal to:

$$\frac{1}{y_3}\left(\frac{C_{m1} + C_{m2}}{y_1} + \frac{C_{m3} + C_{m4}}{y_2}\right)(C_{m1} + C_{m2} + C_{m3} + C_{m4}) \qquad (6\text{-}12)$$

The first term measures the adjusted cost to produce one final unit and the second term is the cost that would have been incurred if the scrap rate were zero.

Cost of Rework
In some cases products are reworked or repaired rather than scrapped. The cost per good unit produced of rework is a function of the defect rate p and the cost of reworking one unit, C_{rework}.

$$pC_{\text{rework}} \qquad (6\text{-}13)$$

The most obvious costs associated with rework are the labor and materials costs. Other costs include:

- **Setup Time.** If a product is built in batches, the processing cost is a function of both the setup costs for each process (per batch) and the actual processing costs (per unit). Some rework policies require that batches be split into two: Good parts continue through processing, while those that fail are sent for rework and then reintroduced into the line as a separate batch. In this case there is twice as much spent on setup costs for every subsequent step. The cost of the extra setup per unit due to a yield issue at station j is a function of the batch yield y_{bj} (what percentage of batches need no rework) at that stage, the size of the batches N_b, the cost of setup C_{si} for each station after the point where the batches are split, and the total number of stations n.

$$\frac{1 - y_{bj}}{N_b} \sum_{i=j+1}^{n} C_{si} \qquad (6\text{-}14)$$

The batch yield is a function of the probability that all parts in a batch will pass quality checks at station j.

$$y_{bj} = (y_j)^{N_b} \qquad (6\text{-}15)$$

- **Expediting Parts.** When parts are reworked, their delivery time is often delayed. If this delays the final delivery of a product, excess costs, C_e, may be incurred if the product needs to be expedited (i.e., shipped express).

$$\frac{1 - y_{bj}}{N_b} C_e \qquad (6\text{-}16)$$

Cost of Lost Capacity

Capacity measures how much product can be produced with the existing equipment. Capacity is a function of the maximum throughput enabled or hindered by equipment, downtime, lost product, and inefficiencies caused by starved or blocked stations.

There are three ways variation can impact production floor capacity, all of which can result in a need for additional equipment or shifts or a reduction in the number of units produced. The cost of capacity can be computed in a number of ways: lost profits, excess overhead per part, extra capacity, and so on. The cost of lost capacity is dependent on market conditions (could you sell the additional products?), the contract structure with your customer (are you fined for late delivery?), and the need for additional capacity (how much excess equipment will you need to buy?).

Capacity can drop for several reasons. First, if parts need to be recycled through the same machinery for rework, the effective capacity drops. Second, the effective cycle time is increased by the scrap rate. For example, if a line produces at a rate of 30 parts per minute and 10 percent of the output is scrapped, the effective rate is 27 parts per minute. Excess scrap at an upstream process can cause a bottleneck and starve a downstream process, reducing the overall line usage. Third, if any machine needs to be stopped to fix quality issues, the entire line may go down, depending on the inventory in the system.

In summary, the production rate of the entire system is a function of the production rates of each subsystem, the buffer sizes between the machines, and the mean time to failure (MTTF) and mean time to repair (MTTR) of the individual stations.[1] If the production rate of any station drops due to a sudden increase in the scrap rate, the average number of parts in the downstream buffers will begin to drop (the average number of parts in the buffer is also a function of the production rates, the buffer size, and the MTTF and MTTR), making the system more sensitive to failures. Stations are more likely to be either starved or blocked, reducing the total throughput of the system. The book by Hopp and Spearman (2000) entitled *Factory Physics* describes how these parameters impact capacity and how to quantify the relationships between them.

Cost of Inventory

Excess inventory may be needed in multiple locations to adjust for scrap, rework, and field issues. First, additional incoming material may be needed to buffer against incoming quality issues. Second, additional inventory may be needed in the buffers between manufacturing stations to reduce the chance of starving any station. Finally, additional finished goods inventory may be needed to ensure a steady stream of finished product to the customer in the event of fluctuations in production quality and production rates. In addition to the capital costs of maintaining inventory, there are other, more intangible costs. For example, excess inventory necessitates excess storage areas. Inventory also makes it more difficult to respond to market changes. For example, one company we worked with made a change to a product to increase customer satisfaction. However, because of the company's significant inventory, the entire existing product had to be reworked at a significant cost. Other costs include making it more difficult to trace defects, excess handling,

---

[1]You will need to run a simulation of production to determine the actual production rates.

and lost inventory. The latter happens more frequently than is expected. In one case, a company lost a part that was 8 feet long—it got put in a corner of the factory until it was needed and was not found for several years.

The cost of inventory is dependent on several factors, the cost of goods sitting in inventory, the current cost of capital, any handling and storage cost per unit produced, and the production rate (the cost of capital and the production rate should use the same time limits). When computing the cost of inventory, only include that inventory that is needed to buffer against variation. Inventory may be held for reasons other than variation, including optimal batch sizes and market demand fluctuations.

Cost of Quality Control Plans

Preventing defects from escaping the factory also costs money. These costs can include:

- **Inspection Costs.** Every time a part is checked, its labor content and cycle time increase. Costs increase more if specialty testing equipment is required. In addition, there is a risk of rejecting good parts or systems.
- **Process Monitoring Costs.** Process monitoring or SPC can be expensive. Time is required to take and enter the measurements, and there is a cost associated with reviewing and acting on the data. Several organizations estimate costs of around $2000 per year per SPC chart.
- **Destructive Testing Costs.** As was pointed out in the section dealing with cost of scrap, organizations will often pull a sample product for destructive testing. Destructive testing costs can include both the labor to conduct the test as well as the lost product and revenue. Often, multiple functional groups (e.g., manufacturing, quality, regulatory, and so on) will separately test product for similar defects. Sharing results and materials between functional groups can reduce testing costs.

Cost of Downgrading

In some cases, product that does not pass final test and cannot be reworked may be downgraded. For example, in the production of central processing units (CPUs), chips that cannot operate reliably at the highest speed may be sold as lower-speed chips. The cost of downgrading has several aspects:

- **Lost Profit.** The downgraded product must be sold at a lower price, but its cost basis is still the same.
- **Lost Capacity.** Often there is limited capacity to produce the higher-grade product. Additional capacity or shifts may be needed to fill the pipeline.

- **Overhead.** The organization must track and store the downgraded product. If there is no market for it, the inventory may need to be written off at some later date.

Cost of Maintenance

Products may require repair and maintenance in the field that is not related to defects introduced in the factory. Damage from uncontrollable sources may require the customer to replace parts within a product—for example, if an automobile is in an accident or a military aircraft comes under fire. If there is significant variation in parts, if parts require rework to install, or if selective assembly is used, just replacing parts may be difficult. There are several costs associated with maintenance, including the cost of the new part, the number of parts needed in inventory, and the labor content. Ideally parts are designed to be fully interchangeable and replaceable (i.e., plug-and-play).

6.2.2. Representation of the Total Cost of Variation

To calculate the total cost of variation, it is necessary to develop a representation scheme capable of integrating the diverse cost data sources. The example of a disposable needle assembly is used to illustrate one such representation method.

Figure 6-7 shows a subassembly that connects a needle with a length of tubing to a luer lock used to connect the needle to another assembly such as an IV bag. The three system KCs are contamination in the fluid path, contamination outside the fluid path, and leaks in the fluid path. The variation flowdown for the needle assembly is shown in Fig. 6-8. The KCs shaded in gray are subject to QC. The flowdown forms the basis for the representation scheme.

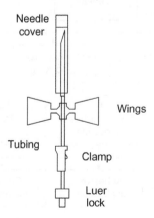

Figure 6-7. Disposable needle assembly.

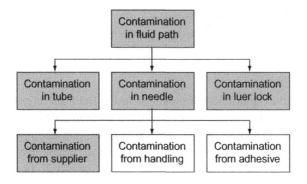

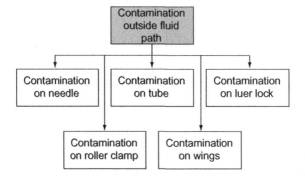

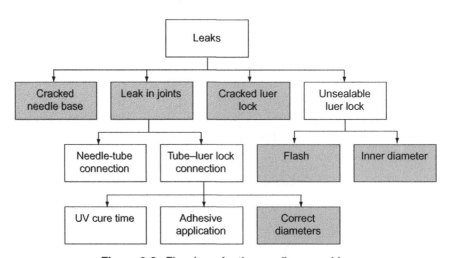

Figure 6-8. Flowdown for the needle assembly.

Table 6-2. Blank cost of variation worksheet

KC	Sub-KCs	Complaint Costs	Scrap/Rework Costs	Quality Control Costs
Contamination in				
fluid path		_____	_____	_____
Tube		_____	_____	_____
Needle		_____	_____	_____
Luer lock		_____	_____	_____
Contamination outside				
fluid path		_____	_____	_____
Needle		_____	_____	_____
Roller clamp		_____		_____
Tube		_____	_____	_____
Luer lock		_____	_____	_____
Wings		_____	_____	_____
Leak in product		_____	_____	_____
Cracked needle base		_____	_____	_____
Cracked luer lock		_____	_____	_____
Unsealable luer lock		_____	_____	_____
Leak in joints		_____	_____	_____
Needle–tubing				
joint		_____	_____	_____
Tube–luer lock joint		_____	_____	_____

After the flowdown is generated, the first step is to determine where the KCs are controlled, measured, and/or inspected. One way to visualize this is to gray out the elements in the flowdown where there is QC (this method will also be used in Chap. 7).

Because all the numerical data will not fit easily on the flowdown chart, it can be entered in a spreadsheet similar to that shown in Table 6-2.[2] The left columns of the sample worksheet contain the specific KCs for the needle assembly, while the right columns contain the relevant cost data. Costs can be assigned to any layer of the variation flowdown. For example, the needle, tube, and luer lock can be individually inspected; in addition, the entire product can be inspected as an assembly. Costs can be calculated on a per unit basis or based on a cost per unit time.

6.2.3. Cost Analysis and Aggregation

There are two parts to cost analysis. First, the group must identify where and when a defect is detected. The location and time of a defect's discovery will

---

[2]For those elements in the flowdown with more than one parent, the KC can be included in multiple locations, or only once with a note indicating that it has more than one parent.

determine the defect's cost. Second, the group must identify where and when the defect was generated. This determines which KC the cost is allocated to. Often the two are not the same: A defect in an upstream station may not be caught until final assembly. For example, contamination may be found on a needle in the final inspection. The cost of scrapping the product is the total cost of the assembly at final inspection, but the cost is allocated to contamination on the needle which was introduced upstream at the supplier site.

The data collected in each category should be entered into the cost worksheet (see Table 6-3). In order to combine the different costs, those incurred inside the factory (cost per good unit times the number of good units produced in a given period) should be measured over a long enough period to get representative values. Combining the complaint costs with costs incurred inside the plant can be tricky because you have to combine real-time data with data on products produced as long as a year ago. Changes in product design or manufacturing conditions can skew the analysis because the

Table 6-3. Cost of variation worksheet

KC	Sub-KCs	Complaint Costs	Scrap/ Rework Costs	Quality Control Costs	Total Cost per KC	Aggregated Cost per System KC	Adjusted Cost per Part KC
Contamination in fluid path		250	200	20	470	830	
	Tube	100	10	20	130		286
	Needle	150	10	40	200		356
	Luer lock	0	20	10	30		186
Contamination outside fluid path		100	40	20	160	315	
	Needle	75			75		107
	Roller clamp	0			0		32
	Tube	0			0		32
	Luer lock	30			30		62
	Wings	50			50		82
Leak in product		200			200	805	
	Cracked needle base	150	20	10	180		230
	Cracked luer lock	50	20	10	80		130
	Unsealable luer lock	110			110		
	Flash		30	10	40		120
	Inner diameter		20	10	30		110
	Leak in joints	30	50	20	100		
	Needle-tubing joint	10			10		85
	Tube–luer lock joint	10	5	40	55		130

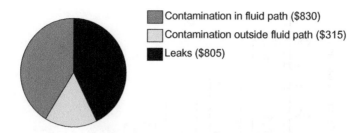

Figure 6-9. Relative contribution of system KCs to total variation costs.

two data sets represent different manufacturing conditions. Unfortunately there is no easy way to address this issue. The most expeditious way is to include current manufacturing data with the most up-to-date complaint/return data and average it across a long enough time period to remove the natural variations.

Once the the data collection is completed, the team can identify the major cost drivers and choose what actions to take. The team should look first at the relative cost of each system KC. For example, the total cost of internal and external contamination and the total cost of leaks are computed by rolling up and aggregating the data for all of the assembly part and process KCs. The contributions of each system KC to total variation cost are shown in Fig. 6-9.

The second part of cost analysis determines the costs of all part and process KCs. Part and process KCs are more difficult to assign because some

---

Do Not Just Do This Once . . .

For products in production, efforts to determine the total cost of quality should not be done just once. There is a significant amount of work that goes into collecting and analyzing this kind of data. When collecting data, thought should be given to how this analysis can be done on a regular basis by updating the data sources. Ideally, the organization should report and track total cost of variation and use it for three purposes. The first is to identify where costs are starting to go up unexpectedly. This analysis can help identify potential high cost problems early. The second is to track how improvement efforts are reducing the overall costs. The third is to continually prioritize where cost improvement efforts should be focused.

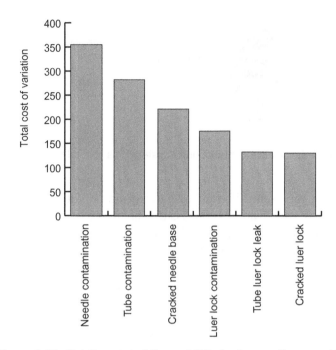

Figure 6-10. Relative cost of the part KCs for the needle assembly.

costs are allocated to the system KC. For example, when the final product is inspected, $200 of scrap is incurred because of contamination somewhere in the product. The data collection system is not specific enough and there is no clear information about where the contamination was found.[3] The simplest way out of this dilemma is to allocate the system KC costs equally to all contributing KCs. For example, the total cost of contamination on the needle is $167—$75 from costs allocated directly to the needle and $32 (one fifth of the cost allocated to the system KC) from the contamination on the whole product. When doing this kind of analysis, many assumptions need to be made. Before taking action on the data, the assumptions should be validated.

Figure 6-10 shows a Pareto diagram of the part KC costs. The first two elements on the list are the particulate contamination. The Pareto diagram can help identify where the efforts in mitigation should be focused and can be used as a baseline against which to compare the results of any improvement efforts.

---

[3]The team should report shortcomings in the data back to the factory and work to improve data collection methods.

6.3. SUMMARY

The assessment of cost is critical to the I-A-M procedures because it ensures that the most costly defects are being worked on first. Calculating the cost is nontrivial: there are many contributors and they each impact cost in a different way. During product development it may not be possible to accurately assess costs because there isn't enough information about the factory and how the product will be manufactured. However, the cost models for similar products can be used for qualitative ranking after the models are modified based on the differences between the new and old products. For existing products, it is necessary to look at the total cost of quality for the present manufacturing processes. A single metric used alone such as customer returns or scrap may give a skewed impression of the major cost drivers.

7

ASSESSMENT OF THE QUALITY CONTROL SYSTEM

The previous two chapters described how to measure the expected cost associated with a system KC's variation. The expected cost is a function of the probability of generating a defect times the cost of repairing or responding to the defect. In addition, there is a third metric: how good the current quality control system is at reducing the number and impact of defects. Quality control increases both labor costs and cycle time to ensure that the product is defect free, hopefully at a net benefit.

The tools outlined in this chapter are most useful for products already in production. In addition, teams can use them to proactively evaluate proposed quality plans before products are released into transition to production. This chapter should be read primarily by those analyzing products already in production. In addition, teams responsible for developing the quality control plan for new products should read this chapter in detail.

When developing a new quality control plan or analyzing one that is currently in place, it is necessary to understand how effective the current quality control process is. There are several measures that define the goodness of a quality control plan:

- Does it help prevent excess variation from impacting the customer or the cost of the product?
- Does it help identify sources and causes of excess variation and defects?
- Does it minimize the value added to the product before excess variation is identified and removed?
- Does it utilize limited quality control resources efficiently?

A company may spend a significant amount of money on implementing strict quality control to reduce the number of defects escaping to the field to acceptable levels. If the efforts are cost effective, the expenditure can be considered justified, but quality control plans may need improvement if one of the four following scenarios applies:

1. Insufficient or ineffective QC allows unacceptable numbers of defective products to reach customers.
2. Extensive and expensive QC is ineffective in reducing the number of defective products leaving the factory.
3. QC is effective in controlling defects, but QC plans are redundant and costs are out of line with results.
4. QC occurs late in the production sequence, increasing scrap and rework costs.

This chapter describes three analysis methods to assess a quality control plan:

- **QC System Maturity.** This determines how likely the QC plan is to detect defects and diagnose their root causes and how effectively the QC system utilizes limited resources.
- **QC Locations in the Manufacturing Process.** The location of an inspection, testing, or SPC determines how much bad product is made before excess variation is detected. The amount of defective product and the value added to a defective product is a function of the distance between the source of excess variation and where the defect is caught.
- **QC Effectiveness Matrix.** This tool categorizes the various QC efforts so that the IPT can understand where QC is effective, ineffective, and/or redundant.

In addition to the data collected for the cost assessment described in Chap. 6, it is necessary to collect all quality control plans for the targeted manufacturing area. All applicable functional groups should provide data including incoming inspection, quality control, regulatory oversight, and manufacturing. When collecting the current quality control plans, the IPT should gather the following information:

- All monitoring, testing, and inspections
- The KCs associated with the QC measurements
- The point in the manufacturing process where defects are introduced (often different from where they are measured or controlled)

- The effectiveness of each quality control effort, including both type I (passes defective parts) and type II (fails acceptable parts) errors.

The QC system maturity, inspection locations in the manufacturing process, and the QC effectiveness matrix are discussed in detail in the following sections.

7.1. QC PLAN MATURITY

QC plans have three measures of maturity:

1. **Detection Capability and Effectiveness.** Ideally, every type of defect can be detected prior to product release from the factory, and the detection rates are acceptably high.
2. **Diagnosis Capability.** Ideally, the cause of the defect can be determined quickly from QC measurements.
3. **Efficient Resource Utilization.** The labor and equipment resources applied to QC are capable of detecting and diagnosing failures but not redundant.

These three maturity levels are discussed in more detail in the following text. They are determined qualitatively by looking at the typical defects and the location and intensity of the quality controls and the defect rates. Detection capability and effectiveness applies both to defects generated by normal variation and the ability to notice increasing defect rates due to special or attributable causes. Diagnosis capability applies to the ability to efficiently diagnose the causes of special or attributable causes of variation. Efficient resource utilization enables the first two with the minimum of resources.

7.1.1. Detection Capability and Effectiveness

Figure 7-1 shows a variation flowdown and the locations of QC plans for the laser collimator example introduced in Chapter 3. The shaded boxes indicate KCs subjected to quality control. In this case, power output defects from excess surface roughness cannot be detected by this quality plan. If the team must wait for feedback from the field, costs can be excessive both because the cost of repairing an unacceptably high defect rate in the field is expensive and because the factory may have built many products with the same defect before it was detected. If there is a safety issue, all defective products in the field may need to be repaired or recalled at a substantial cost.

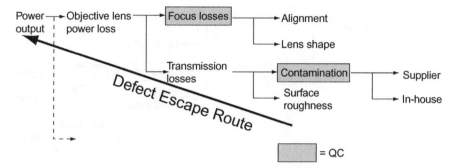

Figure 7-1. Optical system defects not detected by QC.

Rarely do organizations create quality plans that theoretically cannot detect defective products; however, there may be cases where the quality control method can let defective products escape. For example, the product may be subjected to tests that do not accurately reflect actual usage, or a visual inspection is used, which only catches 80 percent of all defects (Juran and Godfrey 1998). The effectiveness of each of the QC inspection points should be evaluated to determine whether defects might still escape.

For the first pass at assessing QC plan maturity, the team should look to see where the QC plan will not find a defect because the quality control is not present. Second, the team should look for where the quality control plan may not find a defect. Looking at customer returns for defects that were theoretically checked can identify inadequate QC. In addition the effectiveness matrix (see Sec. 7.3) provides tools for identifying where both cases may be occurring.

7.1.2. Diagnosis Capability

Figure 7-2 shows the same optical system variation flowdown with an additional QC on the transmission. This QC plan satisfies detection capability and effectiveness—that is, all defects can be caught; however, the QC plan in Figure 7-2 cannot diagnose the source of contamination. If a product fails the transmission check and passes the surface roughness check, it is not clear whether the contamination is internally or externally generated. Although an organization can identify the defect before it reaches the customer, it may take time and effort to collect additional data and locate the source of variation. The setup, measurement, and analysis time required to diagnose the root cause can represent a significant delay in bringing the quality back up to acceptable levels.

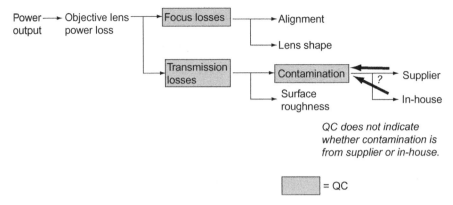

Figure 7-2. Optical system defects not diagnosed by QC.

In some cases it may not be possible to cost-effectively measure enough KCs to enable immediate and accurate diagnosis of the cause of a defect. For example, it may be expensive and time consuming to measure lens shape. In addition, there may be multiple external noise factors that may contribute to power losses that cannot be measured. The production group may decide to not track those parameters that may, but are unlikely to, contribute significantly to variation. If quality degradation does occur in production, variation flowdown can be used to help diagnose the root causes. In addition, the team may decide to put in place contingency measurement plans that can be implemented quickly if an undiagnoseable quality problem occurs.

7.1.3. Efficient Resource Utilization

Figure 7-3 shows the same optical system flowdown, except in this case every KC at every level is subject to QC. Using this QC scheme, the factory is able to rapidly diagnose the root cause of defects. Because of the redundant inspections, the QC plan is not efficient—the team could both detect and diagnose the problem with fewer measurements. Inefficient QC plans have two negative impacts on cost. The first is the cost of the redundant measurements. Some companies have reported that each SPC chart can cost upward of $2000 per year in measurement, labor, and equipment costs. The second negative impact is the reduced focus caused by excessive data. The more data is collected, the harder it is for a team to focus on the critical few pieces of data. It is common to see warehouses of data that has been collected and stored but never analyzed.

In many production environments, a KC will be checked many times for the same defect. For example, it may be checked at a supplier location as

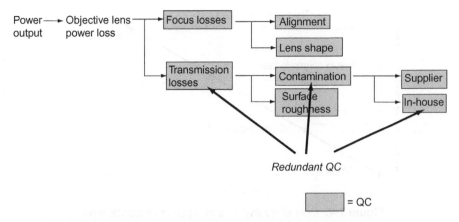

Figure 7-3. Inefficient QC of optical system defects.

well as at incoming inspection, and it may be inspected at different rates—for example, at startup, by sampling, and by 100 percent inspection of finished product. Often these quality control requirements are set by different groups and are not coordinated. By reviewing all QC plans of every internal group and all suppliers and then grouping plans according to the KCs, it is often possible to immediately remove redundant and excessive measurements. In one factory $100,000 of inspection costs per year saved in two days of work by just removing redundant measurements and getting the functional groups to share data.

7.2. QC LOCATION IN THE MANUFACTURING PROCESS

The previous section directs the team to analyze what is being subjected to quality control. In addition, the team needs to think about *where* the quality control is imposed. It is possible to apply QC to the same KC at different points in the manufacturing and assembly process.

Ideally, an organization will detect and fix a source of defects quickly. The farther upstream in the manufacturing process the detection and repair occur, the fewer defective products are built, and the less scrap, rework, and customer dissatisfaction there will be. Detection of defects far downstream of the defect's source or after a product is completed has two major effects on the total number of products built before a defect is fixed. First, the time to identify the existence of a defect is significantly longer because of the processing time and inventory between introduction and detection of the defect. Second, the time to determine the root cause of the defect can also be long, since the team must go back to the production line and analyze the entire

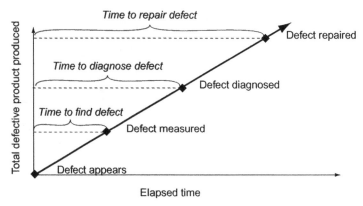

Figure 7-4. Cost of quality as a function of response time.

process upstream of the QC to locate the source of the defect. Figure 7-4 shows the cumulative number of defective products built as a function of the time to find, diagnose, and repair a problem.

Figure 7-5 shows an example of a simple assembly process involving a three-step assembly process chain. The numbers in parentheses indicate the number of units in the buffers. These will be used later in Table 7-1. The first station in the assembly process introduces the original defect in the product; however, it is not until the final assembly that the issue is discovered. If the originating station had discovered the defect, scrapping the part rather than scrapping the final product would save significant cost. For example, if there is a scrap rate of 5 percent, the original part costs 5 cents, and the final product costs $1, moving the QC upstream could reduce the scrap cost for that defect by 80 percent. This savings comes at little expense and often without changing any part of the manufacturing process.

Applying QC as far upstream in the manufacturing process as possible has a secondary benefit. The closer the measurement is to the source of the defect, the more likely it is that the cause of the defect will be detected quickly and fixed. If the upstream supplier or operator cannot detect that they have created a defect, it is virtually impossible for the supplier to improve the quality of the delivered part. A general rule of thumb when doing continual improvement and variation reduction is that *if you can't see it, you can't fix it.*

Often there are several candidate QC points for relocation upstream in the manufacturing process. There are three conditions that should be taken into consideration when prioritizing these candidates:

1. **Defect Rates.** The higher the defect rate, the higher on the list the QC point will be.

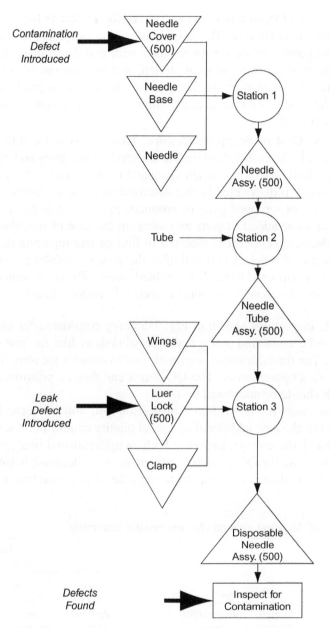

Figure 7-5. Assembly process in which inspection occurs downstream of where defects are introduced.

2. **Number of Products between Where the Defect Is Introduced and Where It Is Caught.** The number of in-process products is a function of the number of steps between the source and detection of defects, the number of products at each station, and the amount of inventory in buffers between stations. The number of in-process products divided by the production rate gives the minimum length of time before defects are detected.

3. **Excess Cost of Scrap or Rework.** Excess cost is the difference between (1) the cost if the defect is detected immediately and the product is repaired or scrapped when the defect is introduced and (2) the cost if the defect is discovered farther downstream in the assembly process. In the case of scrapped parts or products, excess cost is due to the additional value added in parts and labor. In the case of reworked parts or products, excess cost is incurred in finding and repairing defects. For example, if a defect is found after the product is fully assembled, the product will need to be disassembled, increasing labor content and increasing the chance additional defects will be introduced.

The QC inspections shown in Fig. 7-5 being considered for moving upstream can be compared using a simple worksheet like the one shown in Table 7-1. The three measures are multiplied to create a measure of the total cost reduction opportunities. The QC points can then be prioritized to identify which should be addressed first.

In some cases the downstream inspection will still be needed, and the added cost of the new additional upstream quality control point will need to be justified by the savings. Imposing both an upstream and final quality control may increase the QC costs but may create a net decrease in total cost of variation due to decreased scrap costs and a faster response time to defects.

Table 7-1. QC location analysis for the needle assembly

Final Inspection For	Defect Rate		Inventory	Excess Cost per Unit to Remove Defect	Cost Reduction Opportunities	
	Normal Cause	Attributable Cause			Normal Cause (per incident)	Special Cause (per unit)
	p_N	p_s	n_i	C_{excess}	$p_N C_{excess}$	$p_s n_i C_{excess}$
Leaks in luer lock	5%	25%	1000	$4.00	$1000	$0.20
Contamination on needle cover	0.5%	10%	2000	$15.00	$3000	$0.08

7.3. QC EFFECTIVENESS MATRIX

The QC effectiveness matrix method enables a team to assess the effectiveness of the existing quality control plan. Three costs are used in the diagram. The complaint costs (costs of an escaped defect) and the quality control costs (costs of preventing a defect) are used to map each KC as a point on the graph, and the scrap and rework costs are labeled on each point. Based on the relative location of a KC in the graph and the relative scrap/rework cost, it is possible to quickly assess the effectiveness of the existing QC plan and begin to determine the appropriate mitigation strategy. Figure 7-6 shows the 2×2 effectiveness matrix. A KC will fall into one of four quadrants:

1. **Minimal QC, Few Complaints.** KCs that fall in this quadrant can be ignored. Little is being spent and few defects are escaping.
2. **Intensive QC, Few Complaints.** To analyze the KCs in this quadrant, the scrap and rework costs should be considered:
   —**High Scrap/Rework Costs.** If there are high scrap/rework costs, the QC is catching defects before they escape to the field or they are being scraped for the wrong reason. In the first case, the team should look at reducing the scrap rate, then reducing the QC only after

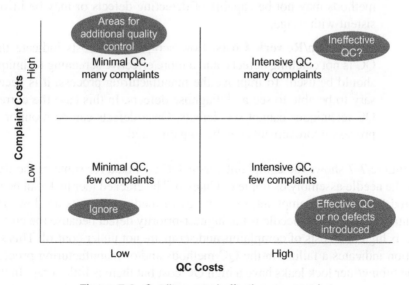

Figure 7-6. Quality control effectiveness matrix.

yields have improved. In the second case, the team can evaluate reducing or removing the quality control.

—**Low Scrap/Rework Costs.** Low scrap/rework costs indicate that QC may be excessive. The team should see whether the QC could be scaled back or whether redundant QC can be removed.

3. **Minimal QC, Many Complaints.** In this case, QC is not catching defects before they leave the factory. It will be necessary to see whether additional QC and/or process improvement efforts will (1) reduce the defect level and (2) prevent existing defects from escaping. The cost and benefit of these two approaches should be evaluated and the best approach taken.

4. **Intensive QC, Many Complaints.** This is the worst of all scenarios. To analyze this situation, the scrap and rework costs should be evaluated, and significant improvements should be made to prevent defects from occurring.

—**High Scrap/Rework Costs.** If there is a high level of scrap/rework, both the production and QC processes must be improved. For example, automatic inspection may need to replace visual inspection. In most cases, visual inspection or other forms of inspection are, at best, 80 percent efficient (Juran and Godfrey, 1998). If there is a high scrap rate and defects are still escaping to the field, adding inspections or making the acceptability criteria more stringent may not help reduce defects escaping the factory because the current QC methods may not be capable of detecting defects or may be inconsistent with usage.

—**Low Scrap/Rework Costs.** Low scrap/rework costs indicate that QC is not catching defects, and a more effective screening technique should be used. To improve the manufacturing process, it is necessary to be able to see and diagnose defects. In this case the current QC techniques cannot see defects. Once defects can be monitored, process improvements can be implemented.

Figure 7-7 shows the complaint versus QC cost data for some of the KCs for the needle assembly example of Chap. 6. The cracked luer lock can be ignored because the complaint, QC, and associated scrap costs are low. The contamination on the needle is the highest-priority defect because the cost of QC is high and costs of complaints and scrap are not under control. This situation indicates a failure in the QC methods and/or manufacturing process. The tubing/luer lock leaks have a high QC cost but there is little scrap. In this case, the team should look at reducing the QC costs for leak checking. Both the cracked needle base and contamination in the tube have high customer

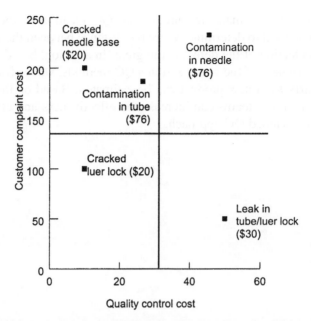

Figure 7-7. Quality control effectiveness matrix for needle assembly.

complaint costs but low quality control costs. In both cases the QC and process control methods need to be addressed.

7.4. SUMMARY

When assessing an existing production line the QC capability and QC costs should be evaluated. While these tools are typically used to retroactively analyze an existing quality control plan, they can also be used during the development of a new plan for a new product to ensure that it is cost effective. This chapter reviews three approaches to assessing the effectiveness and efficiency of current quality control methods.

1. Quality control system maturity
2. QC location in the process
3. QC effectiveness matrix

All of these tools make extensive use of the variation flowdown and process diagrams. Variation flowdown is used to identify where the QC plan is incapable of detecting defects, is incapable of diagnosing the root cause of a defect, or uses too many resources. The QC location in the assembly process can be used to determine where the distance between the source of a defect and detection of the defect is too great. In addition, based on the historical effectiveness of the organization's QC methods, values for the effectiveness matrix for a new product can be estimated. Based on the feedback from these methods, teams can increase the effectiveness and efficiency of the current or planned QC approach.

8

MITIGATION

The goal of variation risk management is to efficiently reduce overall product cost, including costs of supplied materials, in-house manufacturing, QC, scrap and rework, and in-warranty repairs and returns. The purpose of the mitigation phase is to act on the information developed during the identification and assessment phases. The mitigation phase reduces either sources of variation or their impact on system KCs. This chapter reviews the steps involved in the mitigation phase. Chapter sections describe the implementation of mitigation in product development and production and the steps in mitigation. All readers should read the introduction. The body of the chapter which contains the details of mitigation and possible strategies, should be read by those responsible for selecting and executing the mitigation strategies.

Traditionally the QC function was responsible for mitigating the impact of variation. Perhaps more accurately, QC's role was usually limited to testing and inspection to prevent defective product from leaving the factory rather than improving the product. In the last few decades, many organizations have changed their QC philosophies and have made reducing variation and its impact the responsibility of the entire organization and made a broad range of tools available.

It is tempting for organizations to reach the mitigation stage and automatically assume that special quality control (SPC, sampling, or inspection) will be used to reduce the risk. However, in many cases this is not the optimal solution. Mitigation strategies can include any of the following:

- **Design Changes:** Parts and assemblies can be redesigned (or designed originally) to be more robust to variations in parts and manufacturing

137

processes. Adjusting or reassigning tolerances between parts and manufacturing processes constitutes one aspect of design changes.

- **Manufacturing Process Changes:** Alternative processes that introduce less variation can be selected.
- **Manufacturing Process Improvements:** Existing processes can often be improved using techniques such as design of experiments (DOE) and standard operating procedures (SOPs).
- **Manufacturing Process Monitoring:** Tracking processes closely can aid in detecting an increase in variation before KCs go out of tolerance, and can be used to systematically remove special causes of variation.
- **Testing and Inspection:** Increased or improved testing and inspection can catch more defective parts or products.

The best mitigation strategy for each high risk system KC should be selected to appropriately balance the cost of each approach against its benefits. Trade-offs in product development will include "pay me now" for added design costs and possible schedule delays versus "pay me later" for design and process changes (or lower quality) after production has been started.

Figure 8-1 shows the basic steps in the mitigation process. Section 8.1 reviews the differences between applying mitigation procedures during product development and production. Sections 8.2 through 8.5 provide details about each of the steps in the mitigation process.

8.1. MITIGATION DURING PRODUCT DEVELOPMENT AND PRODUCTION

Approaches to reducing the risk of variation-induced defects in products should be data driven and should maximize the profit for the whole organization. The approaches to mitigation are nearly the same for new products under development and for existing products with established manufacturing processes. The major differences are as follows:

- During product development, mitigation tools are applied as early as possible in the project. When choosing the best strategy, cost-benefit trade-offs must be estimated, and the true benefit and impact of efforts is never quantifiable. During product development, the goal is to create a product that is low in cost, has low variation, and is robust to the expected variations in parts and manufacturing processes. The team will typically use design and manufacturing process changes to address high risk KCs.
- During ongoing manufacturing, mitigation tools are applied to actual parts and manufacturing processes. Cost-benefit trade-offs are easier to

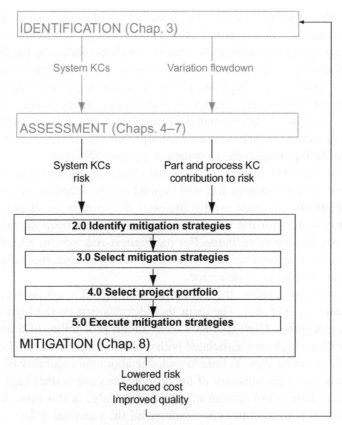

Figure 8-1. Mitigation procedure.

quantify, and the results of actions can be tracked and assessed. For an existing product, the goal is to identify the best opportunities to reduce defect rates and determine the solutions that will cost-effectively address those opportunities. During production, the team will typically use process improvements, monitoring and inspection to address high cost KCs.

8.1.1. Mitigation during Product Development

During product development, the team has the opportunity to significantly impact the cost and quality of a product. If the team waits until production, there is a risk that design and process development will need to be done twice: inadequately the first time and again during production to get it right.

During product development, the team will be simultaneously trading off achieving the target specifications, the cost, and the product quality. Ideally, these discussions should happen early enough in product development to minimize all costs and to reduce the chance of late design changes. If potential weaknesses in a design are identified early enough, it will be possible to employ more proactive mitigation strategies such as modifying the design or the manufacturing concept. This is not to say that a team should use design changes to address every high-risk KC. In some cases, a design change is not feasible due to technology constraints and/or cost, and the responsibility of mitigating the risk will fall on manufacturing. However, it is unacceptable for an IPT to hand over to manufacturing a nonproducible design that could have been fixed during product development at a reasonable cost.

The team should first work to determine where mitigation is required. The effort spent on each system KC will depend on the schedule, the potential impact of efforts, resource availability, and the severity of variation problems. For example, for the lowest-risk system KC, the team may decide to address only the top contributor. For the highest-risk system KC, the team may decide to address three or four top contributors or fundamentally change the way the system KC is delivered.

In the assessment phase, the team prioritized the KCs and put them in the buckets shown in Fig. 8-2. The team should focus first on the high-risk and uncertain categories. High-risk KCs have a high probability of a defect occurring and/or a high cost associated with deviating from nominal. Uncertain KCs are those where there is little knowledge about the capability of the production process or the behavior of the design. They are neither high nor low risk because there is not enough information to judge at this time. More detailed analysis is needed to better understand the potential defect rates and costs of the KCs in this bucket. Another option for the KCs in the uncertain bucket is to implement a change to reduce the risk without quantifying the base line risk, however, there is always a chance the team may waste effort addressing what turn out to be low priority KCs.

The implementation team will be faced with many challenges when applying variation risk management during product development. First, it is not 100 percent certain that the mitigation project will have a beneficial impact on the final cost and product quality. In addition, the mitigation strategies chosen—especially those involving a design or process change—can increase development costs and delay the schedule. When choosing a mitigation strategy, the team should carefully weigh these trade-offs.

8.1.2. Mitigation during Production

During production, the uncertainty about the true costs and benefits of mitigation strategies is smaller, because the cost and impact of most mitigation strategy can be quantified. The biggest issues in applying mitigation during production are the prioritization, scheduling, and execution of improvement

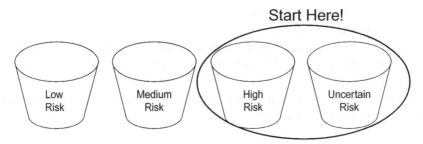

Figure 8-2. System KCs bucketed according to risk.

efforts. Most organizations have large lists of projects under way that "should be finished soon." (One extreme example was a company with 10 engineers who had a list of 100 ongoing projects.) When a problem is detected, the temptation almost always is to create a new project. If there is no systematic prioritization of projects, the newer projects get the most attention until another "fire" starts. Fire fighting leads to long project execution times, incomplete projects, and a tired and frustrated team. (See the text box in Chap. 2 entitled "Fire Fighting versus Focused Problem Solving.")

8.2. IDENTIFYING MITIGATION STRATEGIES

The first step in mitigation is to identify possible mitigation strategies for each high-risk system KC. During the assessment phase, the IPT works to assess risk and identify the major contributors and will use this information as a starting point for the mitigation phase. Each functional group should apply its own expertise and participate actively in identifying possible mitigation strategies. For example, the engineering group may have a clever way of changing the design, or the manufacturing group may propose a new process.

The following sections describe the five options for mitigating risk or reducing costs. In some cases, teams will decide to use a combination of strategies. The purpose of these sections is to highlight the costs and benefits of each strategy as well as provide lists of issues to think about when evaluating whether or not to use each strategy.

8.2.1. Design Changes

Changing the design can significantly reduce the cost of variation but it can also be the most expensive strategy for existing products. For example, on the Boeing C-17, the landing pod structure contained hundreds of parts and took too many hours to build. By redesigning the structure, the team was able to reduce the build time dramatically and improve quality. However, the re-design effort took significant time and required the company to purchase new tooling. In this case, the design change was well worth the effort as the project was estimated to save $45 million on 88 aircraft (McDonnell Douglas, 1997).

When evaluating a design change as a mitigation strategy, a team should look at costs associated with:

- **Creating the New Design.** Designing parts and making drawing changes can require significant engineering resources.
- **Validating the New Design through Testing and/or Simulation.** For regulated industries, the regulators, e.g., the Federal Aviation Administration (FAA) and the Food and Drug Administration (FDA), may require specific and expensive validation or qualification tests.
- **Building New Tooling.** Acquiring new tooling may require a significant lead time and can involve significant costs.
- **Integrating the New Design into the Existing Production System.** In some cases, it is not possible to stop the line for any significant time. As a result, a "running change" may be required in which the new tooling and design are slotted in with minimal impact to the production schedule. A running change involves careful planning and added QC to ensure that the production line is not impacted by unexpected quality issues.

Robust design is the term used to describe a set of methodologies used to ensure that a product is not sensitive to variations in its parts, manufacturing processes, or external noise factors. Taguchi (Fowlkes and Creveling, 1995; Taguchi, 1992) outlines three types of robust design: concept design, parameter design, and tolerance design.

Concept Design

Changing a design concept involves making a fundamental change to the way a product works or is built. Concept design options can include:

- **New Technology.** New technology is the application of a new way of achieving the design objectives. One example is using active noise cancellation to reduce noise levels. Another example of this is controlling a motion using an electronic rather than a mechanical system.
- **Modular to Integral versus Integral to Modular.** An integral design combines multiple functions into a single part. This can improve quality by reducing the number of processes and interfaces. A modular design breaks the individual functions into separate parts, which can im-

prove quality through use of more standard components, reduction in part complexity, and easier testing.

- **Tuning a Product.** In some cases, it will not be possible to reduce variation introduced by a process. A product can have tuning added (or taken out) to improve quality. Tuning, also called *in-process adjustment,* can include such techniques as the use of shims or potentiometers.

- **Assembly Process.** A simple example of a robust concept design appears frequently in the aerospace and automotive industries. When two panels are joined, the overall dimensions can be difficult to control. There are two ways of joining two panels (Fig. 8-3). The butt joint sets the overall dimensions based on the dimensions of the two incoming parts, while the slip joint permits the overall dimensions to be set by the fixture used to hold the two parts. Based on the relative variations introduced by parts and/or the fixture, one of the two assembly processes will will result in less overall variation.

- **Indexing and Datuming.** In assembled products, parts are often located relative to each other by index points (pins in holes) or by surfaces using either part-to-part assembly (also called determinate assembly) or fixtures to locate parts. Either way, the method for datuming parts can significantly impact the amount of variation in the final assembly. For example, Fig. 8-4 shows the same assembly with different datuming structures. In the first, datums are consistent with the goal of the final assembly. Variation in the total length is only a function of variation in L_1 and L_2. In the second assembly, using the inside edge as the datum means that the total length is a function of L_3, L_4, L_5, and L_6. The total variation is likely to be larger in this case.

Parameter Design

Another way to improve a design is to leave a product's fundamental structure and geometry alone but change the individual dimensions and settings.

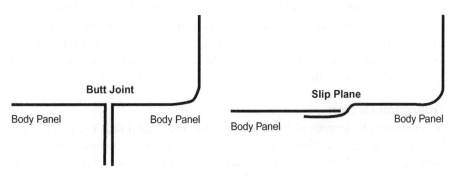

Figure 8-3. Butt joint versus slip plane.

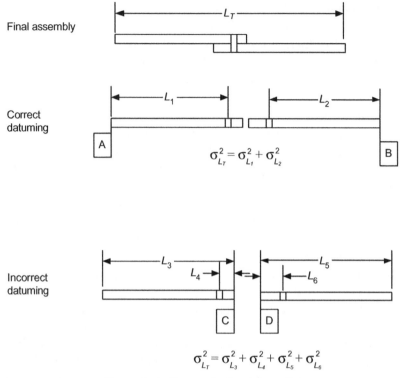

Figure 8-4. Two methods of datuming.

Taguchi and other proponents of design of experiments have provided a huge literature base on how this can be done. Design of experiments is a method used to optimize processes and designs. A minimum set of experiments is done using selected combinations of settings and dimensions. The result of the experiment are the settings that optimize performance and/or minimizes variation. A number of experts have written books on the subject, including Phadke, Taguchi, and Box et al. (Box et al., 1978; Phadke, 1989; Taguchi, 1992). For example, a NVH engineer may determine the optimal amount of insulation to reduce noise while minimizing weight. When executing a design of experiments, it is very important to ensure that the experiment is statistically valid. Experts in the area should be consulted before setting up an experiment to ensure that the results will be valid and useful.

Tolerance Design
A design can be improved by changing tolerances (and thus the process selection and process settings) specified on the drawings. In some cases tolerance design involves reallocating tolerances (reducing the tolerance on one part and increasing it on another). In other cases it involves tightening tolerances to meet performance specifications.

Allocating tolerances can be more of an art than a science. Some organizations get too caught up in the importance of tolerances and fail to remember that tolerance allocation is merely the budgeting of the allowed variation in the customer requirement to the individual features manufactured by the factory. Tolerances can be allocated in any number of ways for the same product; each tolerance scheme will result in different manufacturing processes being selected. Tolerance budgeting can be viewed the same way as financial budgeting. At the end of the day, the balance in a bank account is a function of how much was spent; but a budget allows you to break down spending into categories that are easily tracked. However, setting a budget does not guarantee that you will end up with a positive balance in your bank account if the budget is unreasonable.

Tolerance design works to best match the needs of the product to the current capability of the in-house or supplier processes. As the team learns more about the design and the manufacturing processes, tolerances can be tightened in areas where process capability is better and loosened in other areas.

8.2.2. Manufacturing Process Changes

An organization may decide to change the manufacturing process by which a part is produced in order to reduce variation. An example of this is the move in the aerospace industry from near-net-shape machining to machining from a slab of material. For large structures, especially those used in the aircraft industry, the typical method for machining parts was to forge a piece of aluminum to a near net shape. The part was then machined out of this structure. This approach had several shortcomings, the most important of which, from a quality standpoint, was part distortion. Because of residual stresses in the forging, parts would often warp during machining or heat treatment. In addition, there was a long lead time for parts and any design changes were very expensive. Recently there has been a move toward machining the same parts out of large slabs of raw stock that has been stress-relieved. This approach made possible by a technology improvement in high-speed machining, involves a large amount of "hogging" or rapid removal of material, more material waste, and a longer machining time. However, it does reduce past variation and increases flexibility.

In addition to changing fabrication processes, it is possible to adjust assembly methods to increase quality. Options include selective assembly, in-process adjustments, or fixtures. For each proposed process change, the team will need to evaluate the trade-offs between the improved quality and the increased labor and materials costs. When choosing a new process, the team should consider the following issues:

- What are the recurring and nonrecurring costs associated with the new process?

- Can this new process be used across multiple product lines and/or increase the range of products that can be produced? Will this improve the company's competitive edge?
- Will design changes be required to account for the new process?
- What is the lead time for bringing in new equipment and validating it?

Selective Assembly

Assemblies are usually put together with parts randomly selected from bins. The resultant variation of the assembly can be computed using the RSS method or a Monte Carlo simulation (Chapter 5). Variation in the assembly is usually larger than variation in the individual parts because there is always a chance that variations will combine in a deleterious way.

In selective assembly, parts are matched to offset variations (Boyer and Nazemetz, 1985; Kern, 2003). This is done by measuring all parts and sorting them into bins of similarly sized parts. When designing a selective assembly, the process engineer must choose (1) the number of bins and (2) the methods of binning. Typically one of two approaches to binning is used: equal width and equal area. In the equal width approach, the ranges of part sizes are identical (i.e., 0.01–0.02, 0.02–0.03, and 0.03–0.04). In the equal area approach, the ranges are selected to ensure that there are an equal number of parts in each bin and are based on the standard deviation and mean of the dimension of interest. Assembled parts are selected from matching bins.

Figure 8-5 shows a simple example. The outer cylinder has an inner diameter of 5 and a standard deviation of 0.01. The inner cylinder has a diameter of 4.95 and a standard deviation of 0.01. If these are assembled with randomly selected parts, the gap will have a mean of 0.05 and a standard deviation of 0.014.

However, if the production line measures each part and groups them into one of several bins and then assembles them, variation is reduced significantly. If the parts are assembled using an equal width approach to selective assembly, the gap mean is still the same but the standard deviation is 0.007, one-half that of the random assembly. This benefit does not come without a price—measuring and storing enough parts to effectively match parts is expensive. In addition, it is rarely possible to fully automate the sorting process, so throughputs will be slower and labor content higher.

In-Process Adjustments

In some cases it is not possible to manufacture a product without adjusting subsystems during assembly. Two classic examples of in-process adjustments include shims and tuning. Both are used when it is not possible to manufacture a product to target on the first try.

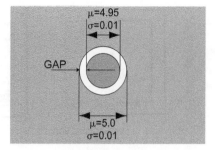

Designed pin in hole

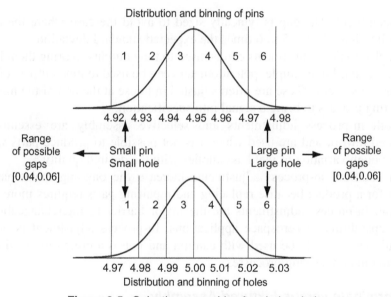

Figure 8-5. Selective assembly of a pin in a hole.

Shims are often used in aerospace applications where (1) a tight fit between two parts is required and (2) it is not possible to guarantee that two
mating parts will not interfere with each other. For example, when a wing is
mounted on an aircraft, the wing essentially slides into the centerbox of the
aircraft. The fit between the two must be tight, and the structures have little
compliance. Figure 8-6 shows a simplified example to demonstrate shimming. The two structures should fit with zero clearance between them. In the
best case, the assembly would be designed with zero gap between the two
parts; however, if the inside part is a bit too big or the outside part is too
small, the two parts will interfere with each other and significant rework will
be required. In order to avoid rework, the two parts are designed with a gap

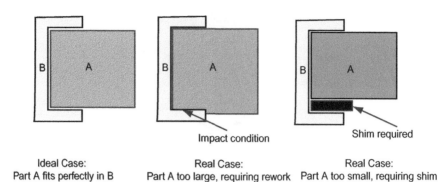

Figure 8-6. Shimming parts to fit.

between them. The gap is typically sized to avoid the case where the two parts interfere—often 5 to 6 times the expected standard deviation.

In electronics, adjustments can be used to tune outputs to bring them into specification. For example, potentiometers can be used to tune outputs of an electronic system. These are often adjusted in house at the end of the manufacturing process to bring a signal into specification.

Both in-process adjustments and selective assembly are essentially planned rework and are used when it is not possible to produce parts with low enough variation to yield assemblies with the desired quality.

In addition, in-process adjustments increase the ongoing maintenance costs for a product because replacing and repairing parts requires more expertise. In-process adjustments are the major barrier to interchangeability and reparability in aerospace applications. In-process adjustment is often useful, but it should be used with caution and not as a crutch to avoid improving manufacturing processes.

Determinate versus Fixtured Assembly

Automotive, aerospace, and naval applications all use fixtures to locate parts relative to each other to aid in assembly. Typically, two skins or a skin and a frame are positioned using hard tooling and are then fastened. In the case of aerospace applications, parts are typically match-drilled and then connected with fasteners (e.g., rivets). In the case of automotive or naval applications, parts are welded together.

In determinate or fixtureless assembly, locating features on the part are used to position parts relative to each other without using hard tooling. Tooling is only used to support the weight of the assembly and minimize deflection. Support tooling is much more flexible to design changes and in its ability to accommodate multiple product configurations. Determinate assembly is being broadly adopted in the aircraft industry because of its cost and flexibility benefits.

Figure 8-7 shows a simple example of determinate versus fixtured assembly. On the top, the hole positions in the two parts are used to set the length. On the bottom, the two parts are loaded in a fixture and joined; in this case, the fixture determines the total length.

There are several benefits to each approach; the IPT should weigh the trade-offs between the two and make the best decision. Determinate assembly has the following benefits:

- **Lower Fixed Costs.** The tooling associated with fixtured assembly tends to be expensive.
- **Easier Design Changes.** Because of the cost and lead time associated with fixed tooling, it can be difficult to make design changes. Determinate assembly gives more flexibility for design changes and product upgrades.
- **Lower Labor Costs.** Much of the assembly time can be eliminated. For example, in aircraft assembly, match-drilling requires parts to be assembled in the fixture, drilled, pulled apart, deburred, reassembled, and fastened. In determinate assembly, the operator only needs to fasten the parts.
- **Less Assembly Variability.** The assembly process used to fixture parts can vary depending on the operators and their training. Standard operating procedures for determinate assembly can be less complex and less subject to interpretation.

Despite all the benefits of determinate assembly, there are several risks associated with it. Primarily, it requires *much* more attention to issues of

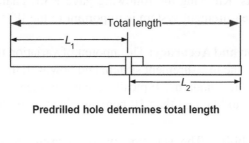

Predrilled hole determines total length

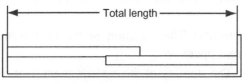

Fixture determines total length

Figure 8-7. Determinate versus fixtured assembly.

variation. Parts that have assembly features pre-drilled can be expensive to rework and require a higher degree of precision than those used in fixtured assembly. Continued attention to the impact of variation on the dimensional integrity is critical to creating designs that achieve variation targets given the current process capability. Because of the challenges of determinate assembly, fixtured assembly will continue to be used in many manufacturing operations.

The benefits of fixtured assembly are as follows:

- **Robustness to Variation.** The fixtured approach allows much more variability in parts.
- **Lower Risk of Rework and Scrap.** Because the fixture allows for significant variability to be absorbed, there is a lower chance of reworking or scrapping parts.

If You Can't See It, You Can't Fix It: The Importance of Gage R&R

When doing any process improvement, process monitoring, testing, or inspecting, the accuracy of the measurements is critical to success. *Gage repeatability and reproducibility* (gage R&R) refers to the ability of a measurement system to accurately determine variation. All too often, measured apparent variation in a KC is actually variation in the measurements. Knowing the following gage R&R characteristics for each KC's measurement system is important to the IPT:

- **Precision and Accuracy:** The amount of variation that can be detected by the measurement system and how well the measurement conforms to standards. Typically, a measurement system's uncertainty should be less than 10 percent of the total allowable tolerance.
- **Repeatability:** The variation in the measurements by a given operator.
- **Reproducibility:** The variation in the measurements taken by two different operators.
- **Stability:** The stability of the measurement over time.

8.2.3. Manufacturing Process Improvements

In many cases, existing processes can be improved enough to satisfy the requirements of a design. As with robust design, a significant amount of literature has been written on improving existing processes. Here is a small subset of a large number of available tools and techniques.

- **Process Analysis.** By mapping manufacturing processes (see Chap. 5), it is possible to identify where unacceptable variations are introduced.
- **Root Cause Analysis.** A variety of techniques, such as fishbone diagrams, process FMEAs, and variation flowdowns can be used to identify what may be causing a process to introduce excess variation (Anderson, 1999; Gano, 1999).
- **Design of Experiments.** By tuning the manufacturing process (speeds, feeds, temperatures, etc.), better performance can be achieved. The optimal settings are chosen by doing a series of experiments to determine performance under different process conditions. Montgomery (2000), Phadke (1989), and Ross (1996) are just three of many books that are available on this subject.
- **Regular or Proactive Maintenance.** Simple techniques such as cleaning machines regularly or changing cutting tools before their performance degrades can improve long-term capability (Gross, 2002).
- **Standard Operating Procedures.** In some cases, different operators will build the same product, operate the same machine, or perform the same process in different ways. Creating standard operating procedures and ensuring operator compliance can remove some variation.
- **Poka-Yoke.** Another name for poka-yoke is error-proofing. By incorporating methods to prevent assembly or process errors, sources of normal variation as well as attributable causes can be removed. For example, a torque-limiting wrench with a feedback light may be used to ensure that a bolt is tightened neither too much nor too little (Shimbun, 1989; Shingo, 1986).

8.2.4. Monitoring and Controlling Manufacturing Processes

Every process contains a natural amount of variation. Both the mean and the standard deviation of a part's characteristic can either drift (due to wear in machines or other factors) or undergo sudden deviations from normal capability. To track the amount of variation in a process, a number of charting techniques, broadly termed Statistical Process Control, can be used to follow the performance of a process in real time. When used correctly, these methods can identify process failures before too many defective or substandard products are made. In addition, manufacturing process monitoring can show where improvement efforts are needed to reduce variation. The SPC chart also shows whether a process is maintaining desired capability. There are hundreds of books on the subject, many more than can be listed here; however, Montgomery (1996) provides a good summary of the various charting methods and how to interpret the data to improve a process.

Figure 8-8 shows a simple SPC chart that is tracking the value of a characteristic. The data points can be single measurements or an average of a set of measurements taken at regular time intervals. The upper control limit and lower control limit are determined based on statistics, sampling size, and the mean and standard deviation of the manufacturing process. If the manufacturing process exhibits only normal causes of variation, the measurements should fall between the UCL and LCL. The appearance of a pattern (multiple points on the same side of the mean, a clear trend, a sawtooth pattern, etc.) or values falling outside the control limits are likely to be due to special or attributable causes (Montgomery, 1996). The attributable causes of variation are those that can be traced to a failure in the manufacturing process or incoming materials. For example, the operator might not follow the standard operating procedures, or a tool might break. Other sources of attributable causes of variation include flawed tooling, off-spec incoming materials, shift changes (if different shifts use different procedures), and setup errors. Measurements falling outside the limits should trigger action to identify possible attributable causes of variation and to prevent the failure from occurring again (i.e., using root cause analysis and corrective action). There are a wide

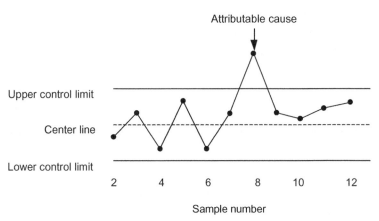

Figure 8-8. Simplified statistical process chart.

range of SPC charts that can be used to track capability including $\bar{x}$, S, and R charts for variable data and control charts for fraction nonconforming for attribute data.

While SPC is an effective tool at controlling existing variation, it does have some limitations. Monitoring the manufacturing process or using statistical process control should only be done to control variation if:

1. The manufacturing process, when stable, is acceptable.
2. When the manufacturing process goes out of control, the capability is unacceptable.

It is important to allocate SPC wisely, because some factories fall into the trap of putting SPC everywhere. When statistical process control becomes "statistical process charting," a lot of data is collected but no action results.

There is no simple solution to determine where to put SPC in a complex product. When determining where to monitor processes, it is vital that the identification and assessment procedures be carried out as an input. When trying to control variation on a system KC, the team can put SPC on the system KC itself or on any of the part- or process KCs that contribute to it. Putting SPC at the top of the variation flowdown gives an accurate picture of the end product quality, but it may be difficult to diagnose the attributable causes of variation. Putting SPC on all manufacturing processes that contribute to the system KC can be expensive and may reduce the focus of the team on the critical few KCs. The holistic view and the assessment of risk and contribution are both inputs into selecting the monitoring strategy. Using an overall holistic view, the IPT must balance several issues when deciding where to apply monitoring:

- **Contributions to Variation.** The manufacturing processes with a larger contribution to variation should be monitored before those with a lower contribution. In addition, processes that are likely to go out of control should take precedence over those that are stable.
- **Cost of Monitoring.** The manufacturing processes that cost less to measure should take precedence.
- **Ability to Take Action.** Some processes will be easier to improve or control than others.

If it is economically feasible, monitoring should be done on the system KC (to confirm control) as well as the critical part and process KCs (to effect control).

8.2.5. Testing and Inspection

The final fallback strategy is to use inspection and/or testing. This mitigation strategy should only be used when process control or a design change is too expensive and/or the cost of a product defect is high. The main benefit of inspection/testing is that it is relatively inexpensive. However, it provides few long-term improvements or benefits to the company.

Organizations should be aware of the limitations of inspection. First, it is often erroneously assumed that 100 percent inspection will catch all potential defects, and as such inspection is often used as a safety net. This assumption can lead to potentially disastrous effects. For example, in the Alaska Airlines Flight 261 crash in January 2000, the inspection of the acme nut thread wear was supposed to catch a potential failure. A measurement system was used to determine the amount of play in the nut. However, the National Transportation Safety Board (NTSB) investigation found that the measurement system, while not a direct contributor to the crash, may not have been accurate enough to detect a potential failure (NTSB, 2002).

Second, go/no go tests applied to variable data can be problematic, especially when products pass inspection but are just inside the allowable variation. It is likely that these products will exhibit problems in the field if conditions are extreme or if any degradation occurs in the product.

Third, tests should be consistent with product usage. For example, in one product, a leak was a critical issue. The leak would not occur unless the product was stressed; however, the product was not leak-tested under stress. In ad-

---

Inspection Pitfalls

One-hundred percent inspection is not a guaranteed way of preventing defects from escaping to the field. Several problems can arise from the use of inspection and go/no go criteria. First, inspection is not always accurate. Visual inspection is assumed to be about 80 percent effective at catching defects (Juran and Godfrey, 1998). This means that one of five products with a defect will escape; a product would need to be inspected three times to ensure an accuracy of over 99 percent. Second, setting inspection criteria can be difficult. We have seen cases where good product is rejected because of incorrect criteria. Finally, inspection is more effective on "goalpost" tolerances, as opposed to cases, where nominal is better.

dition, products that should not leak liquid are often tested with either air or nitrogen. A gas must be used because it would not be possible to remove all liquid prior to shipment, and the product could be contaminated. However, gas exhibits different behavior in the presence of a defect than a liquid does. The team must understand how performance in the presence of a gas corre- lates with that in the presence of a liquid and design the tests appropriately.

Finally, the team should be aware of the impact that inspection will have on the shape of the distribution. Figure 8-9 shows how a normally distributed population of parts can be inspected to select those that fall within the upper and lower specification limits. The truncated distribution will have a higher percentage of products that are near the limits than if the distribution was a normal Gaussian distribution with a C_{pk} of close to 1.33. Table 5-7 provides a quantified example of this effect.

8.3. SELECTING A MITIGATION STRATEGY

This section provides a four-step process for selecting the best mitigation strategy. The input is a list of possible strategies generated by the IPT mem- bers. The four step process is as follows:

1. **Evaluate the Technical Feasibility of the Solutions.** For more radical approaches, teams may need to evaluate the technical and financial fea- sibility of possible mitigation strategies.
2. **Discuss the Organizational Impact of the Solutions.** For each solution, the team should discuss the possible impacts and benefits across each functional group. For example, a design change may require quality

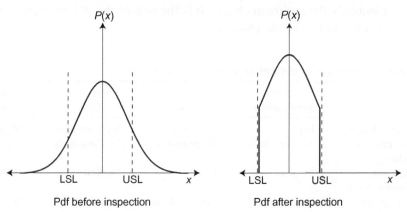

Figure 8-9. Tighter tolerance achieved through inspection.

assurance to develop a new set of testing equipment, which may involve a long lead time. In another case, a manufacturing process change may require the design to be adapted to make it manufacturable on the new equipment. By discussing the impacts across the organization, teams will reduce the chance that they will encounter any unexpected surprises.

3. **Determine the Yield Improvements, Reductions in Recurring and Nonrecurring Costs, and Strategic Impact.** Based on the previous discussion, the cost and benefit of each approach can be evaluated. Each strategy should be evaluated along the four metrics:

   - **Yield Improvements:** Increases in yield have benefits in reducing costs of scrap and rework and in allowing increased capacity. Product design typically has the biggest impact on variable cost reduction and inspections the smallest.

   - **Reductions in Recurring Costs:** These costs are those generated in direct proportion to the number of products made—for example, if choosing inspection there is additional labor to inspect every product.

   - **Reductions in Nonrecurring Costs:** These are the one-time costs of implementing a mitigation strategy, such as a design change or new tooling.

   - **Strategic Impacts:** The long-term benefits of implementing a strategy may include benefits to other new or existing products as well as reducing costs of the targeted product. For example, QC inspections have a low strategic impact because they are not beneficial to future products.

   Table 8-1 shows the typical impact of each of the five mitigation approaches on the four metrics. Because each organization is different, the IPT should fill out this matrix based on its specific operating conditions. There is rarely a case where a single mitigation strategy is obviously the optimum choice. It is the role of the IPT to evaluate the options and make the choices.

Table 8-1. Impacts of mitigation strategies on metrics

	Yield Improvement	Recurring Costs	Nonrecurring Costs	Strategic Impact
Design change	High	Low–high	Medium–high	High
Mfg. process change	Medium–high	Medium–low	Medium–high	High
Mfg. process improvement	Medium	Low	Medium	Medium
Mfg. process monitoring	Low–medium	Low–medium	Low	Medium
Inspection	Low	Low–medium	Low	Low

4. **Rank the Candidate Mitigation Strategies According to Their Potential Overall Cost Savings.** Based on the estimated cost of implementing each strategy, select the portfolio of the highest-priority projects consistent with the budget and resources available. It may be necessary to do more detailed cost-benefit analyses on the specific product areas based on the methods described in Chap. 6.

8.4. SELECTING A PROJECT PORTFOLIO

For products already in production, mitigation strategies are managed as improvement projects. This section mainly applies to products already in production, although the philosophy of good project management should also be applied to products in product development. In product development, using the project approach will be less formal. The team should identify where the major weaknesses in the new design are, determine how they will be addressed during product development, and follow up on these strategies through the gate review process (Chap. 9).

In production, there may be a project to improve a process, reassign tolerances, redesign an assembly, and so on. The efficient and effective selection and execution of the right projects is critical. This section describes how to pick a set of projects and how to best execute them using good project management practices.

It is not possible to implement all mitigation projects. A team is likely neither to implement only one project nor to execute all possible projects; rather, it must select a set of projects that best utilize limited financial, human, and physical resources. The typical metric used to select among projects is return on investment, which measures the ratio of the return from the project against the investment required. However, two additional measures should also be used to select projects. The first is the time required to execute the project. If a project takes a long time to return results but has a high ROI, it may be worth the investment. On the other hand, quick hits that can be executed quickly but have a lower ROI also may be worth working on. The second metric is the project's *probability of failure.* Some projects may have a high potential ROI, but the technology may be unproven and the cost of failure may be large. The team should look at creating a balanced portfolio of projects that span the ROI–time–probability of failure matrix. Other methods for project portfolio management are reviewed in Dickinson et al. (2001).

Figure 8-10 shows the project portfolio matrix. Projects falling in some of the quadrants should not be attempted (low ROI, high risk, long time scale), while others should be done first (high ROI, low risk, short time scale). A set of projects that spans the matrix should be selected. For example, a team may

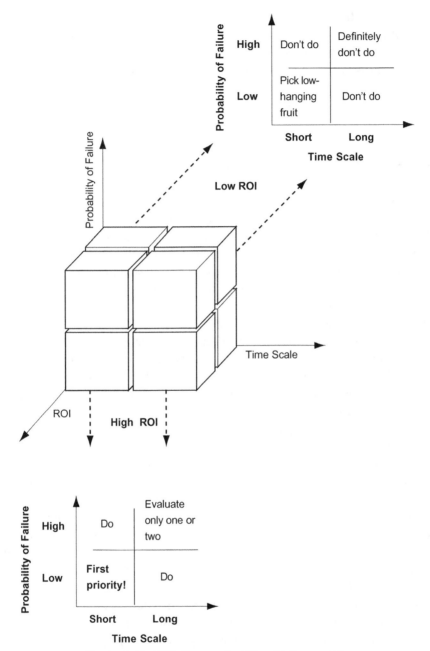

Figure 8-10. ROI–time–probability of failure matrix.

Cost of Parallel versus Serial Projects

Picture a scenario in which 1000 products are being produced per week with 3 projects that will save $5 per product. Assuming each project takes 10 weeks if done serially and 30 weeks if done in parallel, doing the projects serially and getting the returns earlier will increase the savings by $150,000 by obtaining the benefits of project one 20 weeks earlier and project two 10 weeks earlier.

decide to balance the set of projects between those promising low but certain quick returns and those involving higher risk but higher return.

Where possible, it is better to execute the projects do them serially rather than in parallel. Figure 8-11 shows a schematic of two project execution plans. When done in parallel, all projects begin and end together, and, assuming a limited resource pool, the elapsed time to complete each project is greater. No benefits are accrued until time t_3. This is not to say that no projects should be done in parallel, but the team should serialize as much as possible to reduce the elapsed time between project start and project finish. The benefits of performing the projects in series are:

- As soon as a project is successfully completed, the plant begins to accrue benefits. Therefore, if the projects are done in series, each project

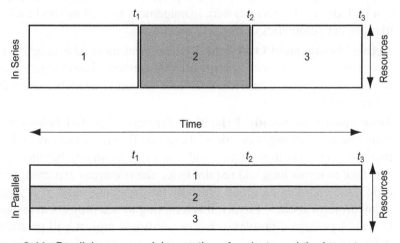

Figure 8-11. Parallel versus serial execution of projects and the impact on completion time for each project.

is completed faster. The plant begins to profit from project 1 much earlier, at time t_1 as opposed to at time t_3.

- Based on the time value of money, money spent later (i.e., postponing projects) is less expensive than money spent today. Because the return does not come until the project is completed, it is best to postpone spending the project money until the project can be completed in a timely manner.
- Completing projects serially allows each project's costs and benefits to be measured more accurately.
- Changes are easier to implement because there is less coordination required among projects and teams have more visibility and accountability.

8.5. EXECUTING MITIGATION STRATEGIES

There are dozens of books on project management. All of them emphasize that projects should be systematically managed and tracked to ensure that they are progressing and that the deliverables will be completed on time. A few key requirements for managing projects that most project management books emphasize are:

- **Clear Project Timelines, Resource Allocation, and Deliverables.** The project leader should be responsible for outlining the project scope and approach to prevent vague projects, unexpected costs, and time overruns.
- **Accurate and Timely Reporting of Current Project Status.** Teams should be required to report current project status regularly. Reporting can include regular reviews with management as well as short and regular project summaries during IPT meetings.
- **Active Management Oversight.** Management must take an active role in overseeing the project's progress. Management should create incentives to keep projects on track and make resources available when they are needed for projects.
- **Maintenance of Steady Priorities.** Projects often fall behind when their emphasis changes or "fires" flare up. Teams will always take on more projects than they should, and it is a rule of thumb that most projects take twice as long and require twice the resources expected. In addition, the engineering group typically has bragging rights if it juggles many projects at once. Finally, it is tempting to add projects in response to real or perceived problems. Every additional project will potentially delay existing projects in the pipeline. A strict protocol for analyzing ex-

isting projects and adding new projects should be implemented. Before adding a project, teams should ask:

—Is this problem so critical that it requires other projects to be put on hold or delayed? How does it compare in benefit and criticality to the top projects already under way?

—What financial, human and physical resources will be needed, and what projects will be delayed?

—What other projects will be impacted by the changes caused by the new project?

Several tools should be used for project management to ensure the timely and successful completion of the project.

- **Schedule.** Each project should have a clear Gantt chart containing a set of tasks, their interdependencies, the resources required, and the major milestones. The timeline should be maintained by a project leader. The IPT lead should be responsible for reviewing the timeline to ensure that it is updated and being followed. The team should avoid creating an overly optimistic schedule and should include reasonable estimates of completion times including possible delays and iterations. The schedule should also not assume that every person will be working on the project full time. Task times should be based on how much time was needed in the past, not how long individuals think a task should take.

- **Project Summary Document.** A project summary document should be maintained containing:

  —**Team Members.** Include a list of all team members, their phone numbers, their e-mail addresses, and their responsibilities in the team.

  —**Project Priorities.** The project rank should be used to distinguish those that are critical from those that are less time sensitive.

  —**Benefits and Costs.** This section should include a measure of return (cost or defect reduction) as well as a budget including resources and equipment.

  —**Project Goal and Description.** This is a simple three-sentence description of the project.

  —**Key Risks and Contingencies.** The team should document possible events that could result in project delay or failure. The team should have a list of contingency plans for each of these possibilities.

  —**Roll-out Plan.** Often teams will spend time on a project investigating a solution but not creating a plan to roll the solution out into the factory. The plan should include a set of tasks with a timeline, estimated costs, and responsibilities for deployment.

—**Current Status.** The status should include current tasks under way, an updated schedule, and any unexpected issues that have arisen.

• **Clear Communication between Project Team Members.** Relatively frequent team meetings should occur to avoid delays, duplication of efforts, and people going off track, and to ensure a clear handoff of responsibilities from task to task. The potential risks should be reviewed on a regular basis to ensure that they are avoided. In addition, potential delays and solutions to delays should be identified.

• **Status Meetings.** The project overview should be updated on a regular basis and reported to the management. During this meeting, project status, completion time, and any potential roadblocks should be discussed.

• **Wrap-up Meeting.** At project completion, the team should evaluate how well the project met its technical, cost, and schedule goals. Any lessons learned should be documented and communicated to other teams.

8.6 SUMMARY

Teams are often tempted to begin with the mitigation phase and start to try to solve problems before the critical issues have been systematically identified and prioritized. For a new product being developed, the team should try to first design the product to be robust to the known sources of variation and, if necessary, make changes to processes. Additional QC should be used only if the simultaneous requirements of quality, cost, and performance cannot be reliably met.

For existing products currently being manufactured, mitigation projects should span the range from quick certain hits to longer-term, high-return but high-risk efforts. The best mitigation strategy should be selected based on the projected benefits, recurring and nonrecurring costs, and the long-term strategic impact.

Fig. 8-12 shows the steps in the mitigation process, its inputs and outputs. Before starting mitigation, the team should understand the following outputs of assessment:

• **System KC Risk.** The relative variation risks of system KCs are used to prioritize which system KCs to focus on. While all system KCs may require mitigation strategies, those with the highest risk may require multiple projects.

• **Part and Process KC Contributors to Risk.** The contribution analysis highlights which part and process KCs contribute the most to the risks

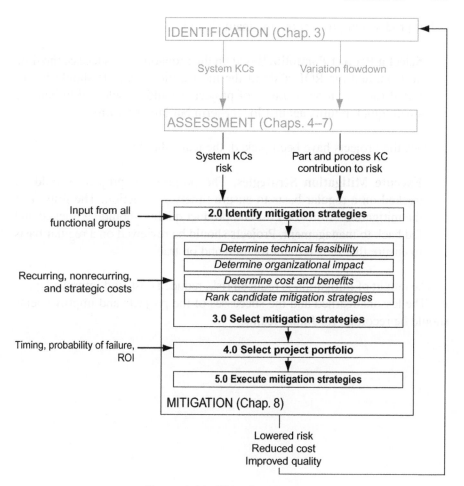

Figure 8-12. Mitigation process.

of excess variation and aids in prioritizing which part and process KCs should be addressed.

Mitigation should start with

- **Identifying Mitigation Strategies.** For each high-risk or uncertain system KC, evaluate possible strategies for reducing variation or its impact, or gather better information about the risk.
- **Selecting Mitigation Strategies.** A decision should be made as to whether to make the change at the system, part, or process level. The mitigation strategies should be evaluated on the basis of their benefits, recurring costs, nonrecurring costs, and strategic impacts.

For products in production, the team should:

- **Select a Project Portfolio.** Based on the probability of success, the time horizon, and the ROI, a set of the most critical projects should be selected for execution. A range of projects should be selected to balance small, quick returns against longer-term but larger returns.

Once the projects have been picked, the team should:

- **Execute Mitigation Strategies.** The progress of projects should be tracked on a regular basis to ensure timely completion. The impact of the mitigation efforts on cost and product quality should be tracked and fed back to management. Projects should be reviewed on a regular basis to ensure that they remain focused and on track.

The output of mitigation is reduced risk, reduced cost, and improved quality. The lessons learned and the successes of the projects and improvements should be recorded for future teams.

9

INTEGRATION OF VARIATION RISK MANAGEMENT WITH PRODUCT DEVELOPMENT

This chapter provides an outline of how to integrate VRM into product development. For proper integration, it is necessary to understand (1) what information is available at each product development stage, (2) how the VRM methodology outputs influence the tasks at each stage and (3) the correct tools to be applied at each stage.

Readers interested in the implementation of variation risk management in product development should spend significant time reading this chapter. Those who are focused on production can skip to the Sec. 9.9.

9.1. BASICS OF PRODUCT DEVELOPMENT

Product development is broadly defined as tasks performed to bring to market a product that satisfies a customer need. The process includes everything from gathering customer requirements through production and shipping. Variation risk management should be treated as an integral part of product development rather than as a separate set of tasks. Starting VRM as early as possible—preferably in the concept development phase—reduces the chances of the IPT either postponing or ignoring variation-related issues. Early application of VRM enables the most cost-effective solutions to be applied to maximizing quality and minimizing costs.

9.1.1. Stage Gate Product Development Process

It is assumed that teams are using a stage gates approach to product development. During each *stage*, a set of specific tasks is performed and specific deliverables are generated. At the end of each stage there is a review process, or *gate*. The criteria for passing to the next stage are clear and are evaluated by the review team at the gate review. If requirements for passing a gate review are not met, one or more of the following actions is taken:

- The project may be deemed technically or economically unfeasible and terminated.
- The IPT may be instructed to redo the present stage or an earlier stage.
- The IPT and the project may be restructured.
- The schedule for the project may be extended.
- Funding may be increased or decreased.
- The IPT may be authorized to proceed to the next stage.

Product development at most companies is divided into several stages (McGrath, 1996; Ulrich and Eppinger, 1995). The following is a typical set of stages:

1. **Requirements Development.** The voice of the customer is understood and translated into requirements.
2. **Concept Development.** Technical solutions are developed and evaluated to determine the best design concept.
3. **Product Architecture Design.** The basic architecture of the product is developed, including how it is subdivided into the major systems and the interfaces between systems. This is typically the responsibility of the systems engineering function.

Challenge the Design

The whole point of variation risk management is to generate a product designed for both the target requirements and the target tolerances at a reasonable cost. The I-A-M procedure challenges the IPTs to identify the most effective strategies to achieve the critical tolerances. It is tempting to constantly postpone addressing the difficult problems, hoping that they will go away. However, the head in the sand approach rarely works and usually results in significant pain and fire fighting later on when changing the design is more difficult and costly.

4. **System Concept Design.** The concepts and approaches for the designs of the major systems are developed by the responsible IPTs. Systems engineering is responsible for how the systems will integrate at the product level.

5. **Detail Design.** Detailed design drawings are made and manufacturing processes are selected or developed for the systems and the product. Dimensions and tolerances for all parts and manufacturing processes are finalized.

6. **Product Testing and Refinement.** The individual systems are reintegrated into a prototype (typically a physical prototype) to evaluate the product performance as a whole. There may be minor design changes at this point.

7. **Transition to Production.** The product is manufactured in small quantities on production equipment. Teams work out any defects in the production methods and the product.

8. **Production.** During full-scale production, the IPTs should work to continually reduce cost and improve quality.

9.1.2. VRM during Product Development

Traditionally, the goal of product development was a product that achieved the target requirements. However, when the effects of variation are taken into

---

When Are You Finished?

Perfection is the enemy of completion. However, the IPT needs to work to reduce the sources and impact of variation as much as possible. Somewhere between these two lies the answer to "how do you know when you are finished." During each stage, the teams should identify what key characteristics are the highest priority and reduce the cost and probability of defects of those high priority items. As high priority issues are addressed and reduced in priority, other key characteristics will surface as the most critical. The easiest completion date is imposed by the hard deadlines related to the gate reviews. In this case, the team is finished when they run out of time. The gate reviews will act as a check to ensure that the yields and quality are sufficiently high to warrant passing the product onto the next stage.

consideration in product development, the scope of the IPT's responsibilities broadens to include ensuring that the product performs at or above expectations and is produced at minimum cost with minimal variation and defect rates. Variation risk management treats meeting tolerances on the product requirements as being as important as meeting the requirements themselves. The IPT is challenged to either trade off among the competing goals of performance, quality, and cost, or, ideally, to find clever solutions to optimize all three goals simultaneously.

Variation risk management is achieved by iterating through identification, assessment, and mitigation at each stage of product development and delivery. Early in the product delivery process, variation risk management will apply broadly to the whole product using qualitative and engineering estimations. At each successive stage the I-A-M procedures will employ increasingly sophisticated and time-consuming techniques, but will be focused on the critical few KCs (Fig. 9-1). At each stage, the IPT will need to start with at least one baseline design, create the variation flowdowns, and assess the quality of that design. Improvements and changes can be made based on that baseline design. Refer to Chaps. 3 through 8 for detailed discussions of the I-A-M procedures.

Figure 9-2 shows an overview of the VRM steps and tools and how they integrate with product development.

9.1.3. Metrics

Metrics are important in that they enable management to track the success of variation risk management efforts. At each gate in product development, the team should measure the progress of the VRM methodology. As the adage goes, "Tell me how I am measured and I will tell you how I will behave."

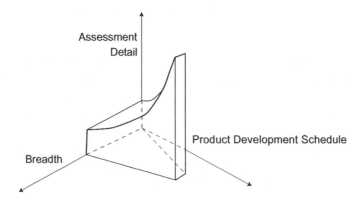

Figure 9-1. Analysis detail and breadth versus time.

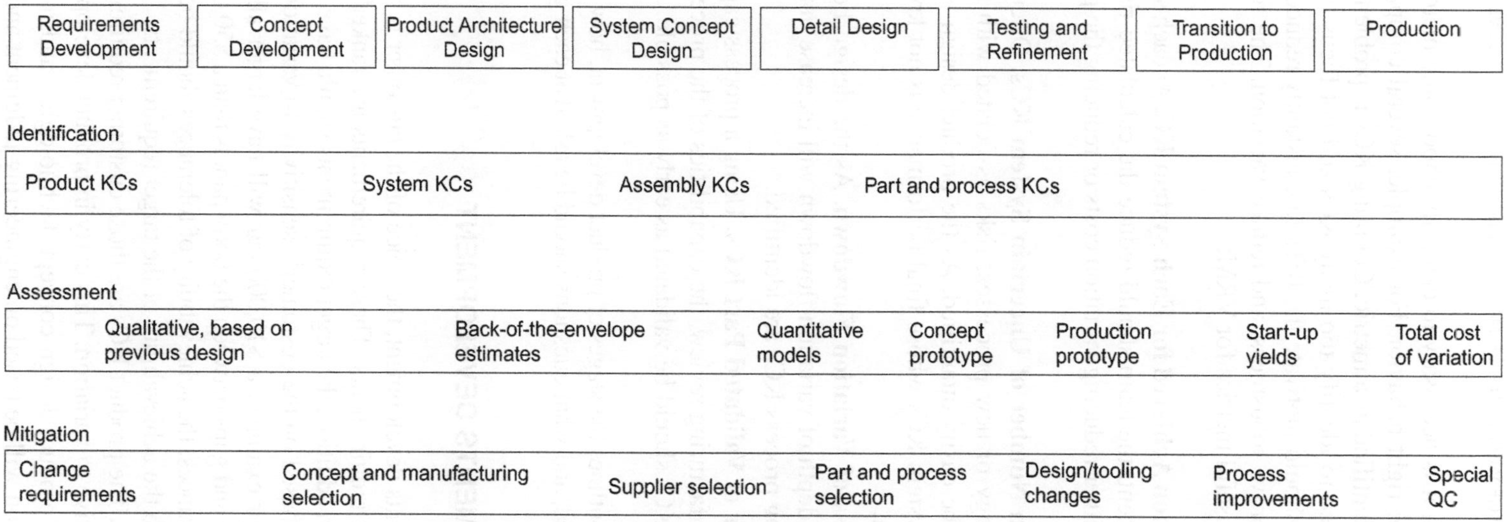

| Requirements Development | Concept Development | Product Architecture Design | System Concept Design | Detail Design | Testing and Refinement | Transition to Production | Production |

Identification

| Product KCs | | System KCs | Assembly KCs | Part and process KCs | | | |

Assessment

| | Qualitative, based on previous design | Back-of-the-envelope estimates | Quantitative models | Concept prototype | Production prototype | Start-up yields | Total cost of variation |

Mitigation

| Change requirements | Concept and manufacturing selection | Supplier selection | | Part and process selection | Design/tooling changes | Process improvements | Special QC |

Figure 9-2. Integration of VRM with product development.

169

With this in mind, it is necessary to carefully choose any metrics to ensure that they drive the right behavior. For example, several companies use the number of KCs identified as a metric. Counting KCs is problematic because the team is tempted to identify too many KCs and not focus on the critical few. The metrics should motivate the IPTs to constantly reduce risk, reduce uncertainties, improve robustness, and reduce variation. The following is a suggested list of useful metrics for VRM:

- **Risk Reduction Achieved for Each System KC.** At each stage of product development, the team should reduce the calculated risks. This can be done by either reducing variation costs or reducing the probability of defects.
- **Reduction in Number of Uncertain System KCs.** Often, because of new technology or new processes, risks associated with some system KCs cannot be easily quantified. As the product design is refined, the number of system KCs whose final performance is not known will become smaller.
- **Completeness of Variation Flowdown.** As the design becomes more detailed, the depth of variation flowdown will increase until the part or manufacturing process KCs are identified.
- **The Number of Validated Part KCs.** Using a process capability database or manufacturing review, the capabilities of the processes generating the part KCs should be validated as early as possible.

The next section outlines the stages of product development, how I-A-M procedures are integrated, and what questions should be asked at each gate review.

9.2. REQUIREMENTS DEVELOPMENT

During requirements development, the voice of the customer is translated into the target product requirements. These requirements are ranked according to customer needs. In addition, the target requirements should have allowable tolerances assigned based on the customer's sensitivity to variation and perception of quality. For example, a $10,000 car will have larger allowable tolerances on the steps and gaps between the body panels than a $50,000 car. When setting target tolerances, the achievability of tolerances should be evaluated simultaneously with the achievability of the target requirements.

Identification. The product KCs are those customer requirements that are likely to be sensitive to variation. This early in product development, little is known about the product design concept, technologies, and manufacturing, so the identification will be based on engineering judgment and previous designs. The product KCs may change over time as better information becomes available.

Assessment. Assessment, like identification, cannot be based on a quantitative analysis of the product at this stage. However, the team can ask the following questions to qualitatively assess the probability of achieving the target tolerances:

- Have we built anything like this before? Are the tolerances significantly different from the previous design? How well did we do on previous designs?
- How does this new product differ from previous designs and how will the difference affect the product KCs?
- What is variation in our competitor's products? Are our target specifications and tolerances similar to theirs?
- Are there any new and unproven technologies in this product that may impact the product KCs?
- Where will defects impact the cost of the product the most?
- What does the customer care the most about?

This discussion should involve the whole group. The individual team members should be prepared with the relevant data and examples to answer the preceding questions. Based on engineering judgment, the highest risk product KCs can be identified. The team should use a qualitative 1-to-10 ranking of both probability of defects and cost of defects to rank the KCs.

Mitigation. This early in product development, few decisions about the technology and product concepts have been made. However, based on the challenge and possible cost associated with meeting target tolerances, requirements can be revisited and modified to better balance the cost-quality trade-offs. Customers should be re-interviewed, requirements analysis revisited, and competitor products analyzed to determine if challenging tolerances truly reflect customer needs.

Gate Review. At the end of the requirements phase, the management team should review requirements to determine if the product should go on to concept development. The review should include questions such as:

- Which product KCs' target tolerances were determined and which were not?
- What is the risk associated with each product KC and how was this determined?
- Have we built a product like this, what were the tolerances, and how well have we achieved them?

- What are our competitors' target tolerances and how well did they achieve them?
- What new technologies and new processes are we planning on using, and what are the risks associated with them (related to yield, rework rates, etc.)?
- How certain is the team that each risk can be mitigated by the design concept? What is the chance that the target tolerances may not be achieved reliably?
- What are the key trade-offs relating to quality (cost versus quality, new technology versus quality)?
- What efforts will be taken to reduce the risk?
- What are the major issues with our current products, and how will these problems be mitigated in this product?

9.3. CONCEPT DEVELOPMENT

During development of the design concept for the product, candidate technical solutions for the product's design and manufacturing methods are proposed and evaluated to determine the best approach. Inputs to the design concept are the technical requirements and product KCs. Based on these, the IPT will work to generate a rough design layout and make key technical choices. These technical choices can include the identification of critical and new technology and manufacturing processes. The technical choices concept selection should take into consideration the ability to achieve the quality targets. Often a preliminary FMEA is performed at this point to identify potential failure modes for the product and its manufacturing processes.

The objective of the VRM methodology during development of the design concept is to ensure that the layout and technical choices are robust to variation. The VRM procedures applied during this phase will be mostly qualitative. Any quantitative analysis will be based on similar products that have been or are being manufactured.

Identification. The product KCs should be evaluated further. Additional product KCs may be identified through FMEAs, technology assessments, manufacturing analyses. Given the basic concept for the product, the team can further detail the variation flowdown. Some system KCs can now be identified. In addition, critical manufacturing processes may be identified and investigated as to their capabilities.

Assessment. Assessment will be based on engineering judgment and estimates of the risks. The previous assessment, done during the requirements generation stage, should be revisited based on the product's concept and the manufacturing choices. The product KCs should be reprioritized based on

---

VRM and Product Platforms

Historically, organizations have designed one product at a time. However, as organizations try to leverage product development costs across multiple products, more are taking a platform approach to product development. A product platform approach enables an organization to simultaneously design products for multiple markets and/or price points. For example, the Black & Decker tool line shared a common motor and battery structure that allowed the company to reduce manufacturing costs and share development of a new motor across several products (Meyer and Lehnerd, 1997). On the other end of the technology spectrum, Lockheed Martin is designing the Joint Strike Fighter, a common fighter platform that can be configured to serve the Navy, Air Force, and Marines. The aircraft can be configured for conventional takeoff and landing, or short takeoff and vertical landing.

Product platforms can reduce product cost in several ways (Meyer and Lehnerd, 1997): *commonality, compatibility,* and *expandability.* Commonality allows multiple products to use the same components. For example, several cars may use the same steering wheel or seat motor. Compatibility allows for the same basic product to have different features and/or qualities of parts. For example, the Sony Walkman can have a significant range of price and feature sets, but the basic product architecture is common across many models. Finally, a product can be expandable—for example, bookshelves whose size and layout can be varied by arranging compatible modules and shelves.

Variation management is one of many issues (e.g., market analysis, product architecture, and costing) that must be addressed when developing a product platform. Variation management is critical when designing the interfaces that enable compatibility, commonality, and expandability in product portfolios. The cost savings accrued through shared development and manufacturing costs can be quickly eaten into by rework and scrap if the interchangeable parts turn out not to be so interchangeable.

---

the updated assessment. Based on the initial design and manufacturing concepts, similar products can be evaluated and used to predict the capability of the new design. The entire team should contribute to the assessment.

Mitigation. The mitigation strategies identified should address the product's design concept and manufacturing choices. At this point in the design there is still a significant amount of flexibility. The team should use the VRM methodology to challenge the design and identify concepts that will reduce the impacts of variation. For example, an organization designing an optical product decided to fundamentally alter the layout of the subsystems rather than buy more expensive components. This decision reduced the rework and assembly time but required rethinking the entire product layout. This effort was possible because the layout had not been finalized and the design was still fluid enough to accommodate the change.

Gate Review. During the review of the product's design concept, the management team should be questioning the IPT to ensure that the design is capable of achieving the performance, quality, and cost targets. Once the product's concept is locked in, changes can be expensive and the ability to reduce sources and impacts of variation is significantly decreased. The questions to be addressed include the following:

- What is the risk associated with each product KC and how was this determined?
- How does the design concept impact the product KCs?
- What were the critical trade-offs between performance, cost, and quality? Are there any clever ways of simultaneously optimizing all three?
- How do the decisions on manufacturing processes impact the risk of each product KC?
- How certain is the IPT that the identified risks can be mitigated by means of good design? What is the chance that target tolerances may not be achieved reliably?
- What efforts will be done during the next phase, system architecture design, to reduce the risks of defects?

9.4. PRODUCT ARCHITECTURE DESIGN

During this phase, the architecture of the product is determined—that is, how the product is subdivided into major systems. Customer requirements and tolerances are allocated to the individual systems, and the interface requirements between the systems are defined. The responsible team at this stage is typically the system engineering function. At this point, the product lead will typically assign IPT leads and staff to the individual system IPTs.

The system architecture decisions are based on a number of criteria including ease of assembly, decoupled design (i.e., minimizing the number and complexity of system-to-system interfaces), expertise of both suppliers and IPTs, and planned product upgrades and changes. The allocation of requirements and tolerances should be done based on previous designs, expectations

Manufacturing Process Development

In some cases, new processes will be developed in parallel with new product development. When developing the processes, the team should focus on several issues including:

- **What Capabilities Do the Processes Need to Achieve?** The target capabilities will be based on the needs of the KCs of the product.

- **How Do We Validate the Capabilities Prior to Production?** Prototype production runs can be used to determine preliminary capability.

- **How Do We Need to Control and Monitor the Process?** Early in the production life of the product, additional mitigation strategies should be included to reduce risk. For example, a plan for inspecting and reworking defective parts or products may be used early in production until the process is stable, defect rates are minimized, and process monitoring can be employed.

of what is technically feasible for each system, and manufacturing capabilities. In addition, systems engineering should select critical system requirements from the complete list of requirements established in the previous phase. The systems engineering group is ultimately responsible for ensuring that tolerances on product KCs are met. However the individual system IPTs should also be involved in determining whether requirements and tolerance targets for each system are feasible.

In some cases, major make-buy decisions are made at this stage. Typically, the suppliers identified are the black-box suppliers who will design and manufacture for selected parts and assemblies. These suppliers act as other system IPTs for the product. (See Chap. 10.) The make-buy decisions are based on technical expertise, cost, schedule, and available resources in house and at the suppliers. The suppliers should incorporate the I-A-M procedures into their product development process. It can be more difficult to ensure the quality of systems delivered by the supplier.[1] Traditionally, quality assurance

---

[1]Some would argue just the opposite. Often it is more difficult to hold internal design teams to quality targets. Because of the contractual relationships between suppliers and customers, there are clearer consequences for missing targets, increasing the likelihood that the targets will be met.

Continuous Processes

This book focuses primarily on discrete manufacturing; however, many of the tools can also be applied to continuous processing. Continuous processing is a class of manufacturing techniques used to make a wide variety of products including metal and plastic sheeting, chemicals, polymers, paper, foodstuffs, and some building materials. Variation management for continuous processes also follows the identification, assessment, and mitigation procedures.

Identification. The system KCs of the product will include the variation-sensitive requirements for the product, and the flowdown will typically include process parameters and input material quality.

Assessment. The quality of the finished product can be difficult to predict a priori because quantitative models are often unavailable. The teams can use historical process capability data and improvement trends to forecast initial and long-term yields. Experimentation will often be needed to understand the relationships between the process variability, variation in starting materials, and process settings and their impact on final product quality. The costs of variation typically include scrap, downgraded product, and reduced production capacity. Often, the majority of the scrap occurs during changeovers and start-ups. Throughput is typically capped by quality because speeding up the line can negatively impact the quality of the product.

Mitigation. The mitigation strategies for continuous processes include process monitoring, process changes, added inspection, and process improvements. Design changes and process change are not usually feasible solutions.

---

is done through carefully written contractual agreements. However, it is better to work with your supplier to jointly optimize the design. This can be done by including the supplier in the in-house IPT and providing the same training and VRM methods to black-box suppliers.

It can be tempting to require tighter tolerances from suppliers than are really necessary in order to maintain spare margin that can be used to absorb excess variation in other systems. The spare margin approach can reduce risk, but it comes at a price—the supplier will inevitably charge for the additional, possibly unnecessary, quality.

Identification. During the identification phase of system architecture design, system KCs are derived from the critical requirements and the product KCs. The systems engineering team should identify critical system-to-system interfaces, potential product failure modes, system KCs from previous designs, and so on to ensure that the system KC list is complete.

Assessment. System KCs and product KCs should be assessed based on engineering judgment, previous designs, and knowledge of the new and existing manufacturing processes and preliminary performance models. Based on the expected variation of the system KCs, the systems engineering team should reassess the entire product for its ability to meet product KC tolerances. The team should also determine which system KCs will have the biggest impact on the product KCs and communicate this to the IPTs.

Mitigation. If there are potential difficulties in meeting the system KC targets, the systems engineering team should consider reallocating tolerances between systems or changing the system architecture design. The team should identify where effort should be expended during the concept product development process to reduce variation or the impact of variation on system KCs.

Gate Review. During the gate review, the system engineering team and system IPTs will be asked questions from the following list by the product team leader.

- How did the system architecture design reduce the risk associated with the product KCs?
- What are the system KCs?
- What is the risk associated with each system KC and why?
    —Have we built technology like this before and what was the experience?
    —How do our competitors perform in these areas?
- What are the highest-risk system KCs and what will be done during the next phase, development of design concepts for each system, to reduce the risks?
- What are the likely key trade-offs within each system?
- Given the risk identified for each system KC during development of the product's design concept, what is the updated risk associated with each product KC?
- What suppliers have been chosen and what are the risks associated with quality for those vendor-supplied systems? How will the risks be mitigated (e.g., teaming, training, contracts, incoming quality control, or maintaining tighter tolerances).

9.5. SYSTEM CONCEPT DESIGN

During development of each system concept design, the layouts, technologies, suppliers, and preliminary manufacturing processes are determined. The output of this phase is a set of rough designs to be detailed in the next stages of product development. One objective of VRM in this phase is to ensure that each system design concept is robust to variation. The tasks during this stage parallel those for the product concept design stage.

Each system IPT has responsibility for the I-A-M procedures for its own system. Overall supervision of the system IPTs and responsibility for system-to-system interfaces resides with the product leader and systems engineering team.

Identification. The variation flowdown generated in the previous phase should be detailed to the subsystem and/or part level. The team should use system design concepts to revisit the system KCs and ensure that the list is complete.

Assessment. At this stage of product development, the assessment done by the system IPTs will be partly qualitative and partly quantitative. The assessment done in the architecture design stage should be revisited based on the system design concept and manufacturing process selection. The system IPTs can do some basic back-of-the-envelope estimates of variation based on previous designs, manufacturing process capability data, rough engineering models, and prototypes. Most of the preliminary assessment will likely be done as a group, but high-risk or uncertain areas will need to be analyzed by individuals outside the group meetings. Conceptual prototypes of the systems and the product can be used to test the design concepts and identify the design parameters that may impact quality. The output is a reprioritization of the system KCs.

Mitigation. Trade-offs between cost, quality, and performance will need to be made to choose the optimal concept. In addition, high-risk system KCs can be identified for attention during the next stage, detail design. The critical suppliers who contribute the most to the risk should be identified and integrated into the VRM methodology. Critical processes that are likely to cause issues in production should also be validated and work should be done to improve them.

Gate Review. During the gate review, each system IPT will be asked questions from the following list by the product team leader.

- What is the risk associated with each system KC and how was this determined?
  —How does the system design concept impact system KCs?
  —What were the critical trade-offs between performance, cost, and quality?

—How do the choices of manufacturing processes impact the risk of each system KC?
- How certain is the team that risks can be mitigated during detail design?
- What is the chance that the target tolerance may not be achieved reliably?
- What efforts will be done in detail design to reduce the risk?
- What suppliers are critical to the delivery of system KCs and what efforts are being made to include suppliers in the design process?

9.6. DETAIL DESIGN

During the detail design stage, the final system tolerances are established; engineering drawings are completed; and materials, suppliers, and manufacturing processes are selected. The IPTs can influence product quality and cost at this stage by means of the appropriate selection of parts, manufacturing processes, and tolerances as well as through appropriate datuming, and assembly sequences. At the end of detailed design, the product's systems are ready for manufacturing prototype and preproduction builds. Geometric dimensioning and tolerancing (GD&T) should be used to specify the part and assembly datums and their KC tolerances. GD&T enables the dimensional requirements of the product to be communicated in a clear and consistent fashion and reduces the chance of misinterpretation in production.

Identification. At the end of this stage, the variation flowdown should be complete. The high-risk system KCs will have been analyzed in more detail than the low-risk system KCs. Branches of variation flowdown with the largest contributors to risk will also be more detailed. The ends of the flowdown branches should be individual process KCs or KCs delivered by internal processes or suppliers.

Assessment. At this point, the IPTs should be using quantitative analysis such as variation modeling for geometric problems or Monte Carlo simulation for nongeometric performance characteristics to quantify risks and identify the major contributors to system KC variations. Only the high-risk or uncertain portions of the variation flowdown should be subjected to this detailed assessment. Historical process capability data should be used (Appendix B) to evaluate a system's producibility. Part KC tolerances should be checked against the process capability database. Process capability data should also be used to populate the quantitative variation models. In addition, manufacturing prototypes should be used to validate the production capability. From these tools, preliminary but believable estimates of yield and cost can be determined.

Mitigation. The mitigation strategies identified will most likely focus on the parts and manufacturing processes selected. In addition, each IPT should identify what problems will need to be addressed in the testing and refinement stage, in transition to production, or in production using either process improvements or special quality control.

Gate Review. During the gate review, each team will need to answer the following questions.

- What system KCs are at risk?
  - —What are the major contributors to each risk?
  - —Can risks be mitigated by modifications of the product's and system's designs, part selection, or alternative production methods?
  - —What will need to be done during product testing and refinement, transition to production, and production to reduce risks of defects?
  - —What were the major trade-offs associated with the high-risk system KCs?
- What process KC risks are still uncertain and need to be validated during the following phase, product testing and refinement?
- How were tolerances set and are the capabilities validated? What process capability data is available to validate the part and process KCs?
- What are the start-up and long-term yield rates expected to be?
  - —How will the long-term yield rates be achieved?
  - —What resources would be needed to increase the start-up yield rates?
- Who are the critical suppliers and how are risks associated with the supplied parts going to be reduced?

9.7. PRODUCT TESTING AND REFINEMENT

During this phase, the designs from the individual system IPTs are integrated into a functioning prototype. Based on the performances of the individual systems and of the entire product, the system designs and their production methods can be modified to improve performance, reduce cost, and improve quality. Typically, parts and assemblies are built using either preproduction or production tooling. Production equipment is tested to evaluate the capabilities of critical processes. The output of the product testing and refinement phase is a production-ready product.

Identification. The variation flowdown should be reevaluated to check for completeness. Any new variation issues should be added to the flowdown.

Assessment. Both the probabilities of defects and the costs of either containing or repairing defects should be reevaluated based on the performance of the prototypes or any detailed performance models. Any KC uncertainties identified during the previous product delivery stages should be reduced and the relative risks of system KCs reevaluated. Manufacturing prototypes can also be used to evaluate and validate initial yield estimates. The assessment phase should identify those KCs for which:

- **Variation in the KC Is Unknown.** These are the KCs that ended up in the uncertain bucket in the earlier assessments, but whose mitigation costs outweighed potential benefits. Process and product measurements may be needed to quantify the actual variation that will be expected in production. This may be difficult if prototype and final manufacturing processes are sufficiently different. These uncertainties will be addressed in subsequent stages.
- **Variation in the KC Needs to Be Reduced.** These are KCs for which the existing variation is known to be too great, and the team has decided to fix the problem through process improvements or quality control.
- **Variation in the KC Is Acceptable, but the Probability and Cost of an Out-of-Control Condition Are High.** This is the case where under normal circumstances the variation in the manufacturing process produces an acceptable product. However, if any attributable causes make the manufacturing process go out of control, the impact on a system KC is significant. In addition, there must be a reasonable likelihood that an out-of-control condition will arise. These will require ongoing monitoring.

Mitigation. The mitigation strategies available in this stage are limited to minor changes in the product, assemblies, parts, or processes or quality control in production. Changes at this stage can be expensive because manufacturing processes and parts have already been purchased and there is very little flexibility in the design.

After the mitigation plans are executed, system KCs should be reranked in order of importance based on their relative risks of defects. Using the KCs selected for mitigation, teams should plan for any additional QC that will be required during the next phase, transition to production. For each system KC the team needs to identify the level, if any, of special quality control that is required. Because system KCs are related to critical system requirements, it may be necessary to maintain some level of quality control on every system KC, independent of the amount of risk. Those KCs with high risk will probably require final product quality checks along with quality control on the part and process KCs that are major contributors to defect rates.

The quality control plan should be formally documented and should ensure that, at the least, defects in all system KCs can be detected. Ideally, the quality control plan should allow for the rapid diagnosis of sources and causes of quality problems and should efficiently use limited quality control resources (Chap. 7). In addition there should be a plan for tracking and acting on QC data.

In addition, the following criteria should be used to select which KCs should be subjected to control and/or monitoring:

- The KC is measurable and controllable. If the system KC cannot be controlled directly and/or cannot be measured easily, the contributors to the system KC should be controlled.
- The cost of measurement and control is reasonable.
- The cost of going out of control is significant.
- The KC's percentage contribution to overall variation is significant.

Gate Review. Before allowing a product to be transitioned to production, the following questions should be answered at the gate review:

- What system KCs are still at risk? What was discovered during testing and refinement?
   —What are the major contributors to the risks?
   —What will need to be done during production to reduce the risks?
- What processes are still uncertain and need to be validated in the next phase?

Truck Door Case

A case based on an experience during the launch (another name for transition to production) of a new model of truck (Leyland, 1997) highlights the importance of actively managing variation issues during transition to production. In this case, the launch team identified a critical "pop-off" problem in which the truck door would spring open when the latch was pulled. Although the problem was identified early, it was not taken seriously until much later because the team assumed the problem would disappear on its own. When the team determined that the problem was of critical importance, it needed to spend significant time collecting data on the relative variation of the door and the body aperture. Had the team already collected the data and had it proactively addressed the issue early in the launch period, significant cost would have been saved.

- What processes will need to be improved during transition to production?
- What KCs will require additional quality control, improvement, and validation?
- What are the expected start-up and long-term yield rates?
  - —How will the long-term yield rates be achieved?
  - —What resources will be needed to increase the start-up yield rates?

9.8. TRANSITION TO PRODUCTION

While the product is being transitioned to manufacturing and brought to full production rate, any delay is expensive. All capital expenditures have been made; there is typically a full staff of production workers being paid; the company has most likely assured the customer of the product's readiness; and the market is waiting. In the worst case, companies spend the transition to production phase working out quality and production issues that could have been solved during product development. In the best case, the organization spends this time working out only those issues that could not have been foreseen, and the team set up a measurement system that can be used to rapidly diagnose critical variation issues as they arise.

During transition to production, it is often necessary to impose special quality control on system KCs and their contributing assembly, part, and process KCs. One output of this phase is the final KC quality control plan to be used in production. The KC quality control plan complements and expands on the normal quality control efforts used by the production teams. The I-A-M procedures during this phase are as follows.

Identification. Identification of all system KCs and variation flowdowns should be complete at this stage. Variation flowdowns should be reviewed before transition to production to ensure completeness and correctness. During transition to production, additional sources of variation (both controllable and uncontrollable) may be discovered.

Assessment. The first action is to validate the capabilities of the manufacturing processes affecting each high-risk system KC and its contributing part and process KCs. Measurements should be taken to determine each process's variability, its impacts on the KCs, and its stability. When determining the capability of a process, the team should assess both the short-term capability and the long-term capability. The individual part tolerances and quality should be assessed as well as the product performance. As has been repeatedly pointed out, achieving the individual part tolerances does not ensure that the product will work as expected because tolerances may be incorrectly set and/or mean shifts may impact the final performance more than expected. At this point, all uncertain KCs should be quantified. In addition, the baseline cost impact of variation should be quantified.

Mitigation. The response to the KCs that have a high risk will be one of two actions: reducing variation or controlling variation.

1. **Reducing Variation.** Excess variation can come in two forms: a shift in the mean value away from the specified value or an overly large standard deviation. If either or both of these aspects of variation occur, a plan is required for reducing variation through process improvement. Once variation is reduced, the IPT should determine whether the KC requires ongoing quality control. Tools to reduce variation can include the manufacturing process improvement procedures described in Chap. 8.

2. **Controlling Variation.** Once variation in a KC is measured and reduced, it may be necessary to impose long-term quality control, which can include inspection to catch defects and/or SPC to prevent degradation in quality. Ideally, the factory should use attribute data either on the part KCs or the manufacturing processes that drive them. The quality team needs to determine which KCs will require ongoing quality control and to evaluate the cost and benefit of imposing that control.

Gate Review. Before launching into full production, the following issues should be addressed:

- What special QC will be needed over the long term?
- What processes still have unsatisfactory capability or are not in control and what is the plan to deal with them?
- What are the current defect levels and what is the plan to reduce them further?
- What is the current cost of variation and how will this be reduced over time?
- What unexpected issues arose and why?

In addition to the I-A-M procedure, several additional tasks need to be executed. These include a plan for customer complaint handling, a wrap-up meeting and documentation of the KC quality control plan.

9.8.1. Handling Customer Complaints

To get good feedback from the field, it is often necessary to proactively work with the customer, directly or through a customer relations/marketing group. To capture useful data that can be interpreted and manipulated easily, a standardized naming scheme and/or coding scheme for defects should be developed prior to the product's release. If the data collection method is not properly designed, the customer complaint data can be vague and at worst useless.

Often, those who collect information on customer complaints do not have a high degree of technical proficiency or were not involved in the development of the product. As a result, they tend to inaccurately describe defects. Complaint handlers often assign causes to problems, even if they do not have the expertise to do so, rather than just reporting the symptoms. Creating standardized terminology and descriptions for defects can make the data collection and analysis much simpler, more accurate, and more meaningful.

There are a number of ways to describe customer complaints. Most companies rely on text descriptions; however, these are hard to formalize and analyze because the data is not quantified, and the text can be interpreted in multiple ways. For example, in one company the term *broken* was used to describe several types of defects in the same part. One way to avoid this confusion is to create a numbering and indexing scheme for encoding the relevant data about a customer complaint. The following is one method that has proven applicable to a wide range of products.

A description of a defect or customer complaint has several components:

- **Part.** Where in the product did the defect occur? By using drawings and/or parts lists, it is possible to ensure that parts have consistent names or numbers and that sufficient detail is given.
- **Defect.** What type of defect occurred? Descriptions of defects can be accompanied by photographs and detailed text descriptions to ensure consistency.
- **Time.** At what point in the product's use did the defect occur? This is critical for understanding the effects and costs of a defect.
- **Use.** Under what conditions of handling or use did the defect appear?
- **Condition of Operation.** Was there a non-standard condition such as the product being dropped, lightning strike, or shipping damage?
- **Batch.** What is the serial number, lot number, or batch number of the product or part? For example, aircraft engine blades must be traceable back to the lot of material and the date and place of manufacture.
- **Date of Occurrence.** This is important for tracking when the defect occurred in case there are seasonal trends that may be affecting the defect rates.

For each descriptor, a hierarchical list of coded responses to each component should be provided. The data collector can add additional textual descriptions if needed. There are several benefits to this approach. First, it eliminates confusion about terms or part descriptions. Second, the data analysis is much easier. Finally, having a coding scheme enables the data collection to be put on a shared platform such as the Internet.

Table 9-1 shows a sample coding scheme for the disposable needle assembly used as the example in the previous chapters. If a user found a black particulate on the needle after the product was opened but before the kit was used on a patient, the complaint would be coded P1.2 D2.1 T2.0.

9.8.2. Wrap-Up

After product development is completed and manufacturing has begun, the VRM methodology and its results should be reviewed. A short wrap-up meeting should be held and the following questions answered:

- Did the variation flowdown accurately capture the critical variation-sensitive requirements? What critical items were missed, and what was included that was unimportant?
- Did the assessments correctly identify the highest-risk KCs?
- Were the correct mitigation efforts executed?

The team should discuss any issues that were missed or any suboptimal decisions. Changes to the VRM methodology should be suggested. Finally, an information packet containing the flowdowns, assessments, mitigation projects, and summaries should be put together for reuse by the next team to develop a similar product.

9.8.3. Documenting the Key Characteristic Plan

The team must decide how to document the KCs for production. The choice is highly dependent on the existing documentation processes and internal

Table 9-1. Complaint coding scheme

Part Hierarchy	Defect Hierarchy	Time Hierarchy
P1.0 Needle assembly	D1.0 Leak	T1.0 Before opening
P1.1 Connector	D1.1 Slow leak <0.01 mL/min	T2.0 After opening
P1.2 Needle	D1.2 Fast leak >0.01 mL/min	T3.0 Prep for use on
P2.0 Tubing	D2.0 Contamination	patient
P2.1 Inside of tubing	D2.1 Black	T4.0 On patient
P2.2 Outside of tubing	D2.2 Hair	T5.0 After removal
P3.0 Cannula	D2.3 White	

company standards. The team must determine if the KCs are maintained on the drawings, in a separate QC plan or both.

Drawing

If KCs are put on the drawing, the implications of this action should be understood. One team caused significant confusion because it did not think about the implication of documenting the KCs on drawings. The design group's definition of KCs stated that the designer should identify all features where variation *could* be a problem. However, the material and supplier management group was working with a different set of assumptions. In its documentation, every key characteristic on a drawing *required* a special quality control plan. As a result, the suppliers were inundated with requirements for special quality control, many of which were not actually necessary.

There are several major benefits of maintaining the KCs on the drawings. The marking permanently identifies those features and/or dimensions as critical to product quality and can be used by later design teams if a redesign is required. However, the KCs on the drawings increase the overhead associated with drawing changes. Adding or removing a KC can be expensive.

Typically, a drawing consists of the isometric views of a single part. Tolerances and manufacturing processes are also included to guide the production of that part. However, the typical drawing does not include assembly information. Many companies are now using integrated product and process design (IPPD) drawings that describe how the individual parts are assembled. The KCs are often called out and explained on the IPPD drawings which have the benefit of showing information about the rationale of the KCs and how the product functions as a system. In addition, this type of drawing provides better information to the factory about how the product should be manufactured.

Quality Control Plan

A separate quality control requirements document (see Table 9-2) should be developed that identifies all KCs and the quality assurance actions being taken. The benefit of this approach is that it is flexible and can be modified

Table 9-2. KC quality plan documentation

KC ID	KC Name	Status			Action (Process Capability Study, SPC, Inspection, Manufacturing Process Improvement)	Description of Action
		Process Unknown	Process Not Capable	Unstable Process		

without altering the drawings. This document will track the current state of the KCs and the actions taken. The document should be updated continually throughout the product's life. If a supplier produces the part or assembly, the supplier should be responsible for the document. The document should include each KC with its ID and name. The status of the KC (i.e., whether capability studies, manufacturing process improvement, or continual process monitoring is required), the type of action taken, and a description of the action should be included. For example, the team may decide to apply an SPC chart to an unstable process. The SAE AS9103 standard includes the standard used by the aerospace community.

9.9. PRODUCTION

Using variation risk management does not end with transition to production. The organization needs to work to continually reduce costs, streamline processes, and improve quality throughout the product's life. Any cost savings will directly benefit profitability, especially in the face of competitive pressure to reduce price. Chapters 3 through 8 describe in detail how to identify, assess, and mitigate the impacts of variation on existing products in production.

In addition to the VRM process the team should execute the tasks described in the following sections.

9.9.1. Continually Monitor Total Cost of Variation

Variation risk management during production should include computing the current total cost of variation costs on a regular basis to provide a metric for determining the benefits of any mitigation strategies.

Using the techniques described in Chaps. 3 through 8, the top contributors to quality costs can be identified, projects with the highest ROI can be selected, and mitigation strategies can be executed effectively. Typically, the QC group is responsible for maintaining this metric and taking action when variation costs are starting to increase.

Periodically, the current status of the product, the total cost of variation, and the current cost reduction, design change, and process improvement projects should be reviewed to ensure continual improvement. Questions that should be asked include:

- What is the current total cost of variation?
- How much has the cost due to variation dropped or increased since the product was introduced?
- What are the major contributors to the cost and what are the major sources of customer complaints?

- What cost reduction, performance, or process improvement projects are ongoing and what is their likely impact?
- Where have changes to the design or production positively or negatively impacted product quality and/or cost? Did projects have the expected benefit?

9.9.2. Track Customer Complaint Data

Using the customer complaint coding scheme described in Sec. 9.8.1, the customer complaint data should be recorded and periodically analyzed. Using the coding scheme, it is relatively simple to determine how many complaints are due to a part or how many defects were detected at a certain point in the product's use.

If the complaint data is well organized, specific defects can be tracked over time to identify trends in the frequency of complaints (Fig. 9-3). Defect rates can be plotted to determine whether the frequency is changing. A statistical approach should be used to understand what types of defects are really increasing or just exhibiting normal variation. An attribute data chart can be used to track the current rate of defects. Based on the sample size (i.e., assuming a standard usage rate) and the expected defect rate, the upper and lower control limits can be set (Montgomery, 1996). Action should be triggered if the graph shows a statistically significant increase in the number of complaints. Improvements in customer satisfaction can also be determined and the control limits can be changed to reflect the decreased defect rates.

Complaints can be tracked based on a number of data aggregation methods, each of which has its own shortcomings:

- **By Date Reported.** The date reported does not give information about usage patterns or when the defect occurred. However, the benefit of this approach is that there is no need to debate whether all data is in.

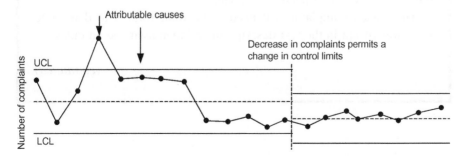

Figure 9-3. Statistical approach to complaints.

Reactive Design of Complaint Systems

If a customer complaint handing system is not designed proactively either before or during transition to production, it may be necessary to redesign existing data collection systems. The complaint and warranty data is typically collected and maintained by a marketing, quality, or warranty group. The team redesigning the complaint tracking should ask the following questions to start:

- Who collects owns the data on customer complaints/returns?
- What format is the data in?
- How was the data collected, and how should the data be interpreted?
- What is the cost associated with a return or complaint?
- What is the response to a customer complaint?
- Are some of the defective products returned for in-house failure analysis, who does the analysis and how, and how are the results recorded and communicated to the IPT?

Unstructured complaint systems usually collect text descriptions of defects. Some complaint reports will contain detailed explanations that include information about the exact location of a problem and some complaints will be general. However, the text descriptions are often subject to multiple interpretations, and the same defect type may be described in several different ways. For example, a leak may be described as a tear, drip, spill, leak, or separation. In other cases, a common term can be used to describe multiple defects. For example, the term *separated* can be used to describe a missassembly or a leak.

The team should generate an indexing scheme for the complaints based on the variation flowdown, and the typical complaints, and the types of analysis that need to be done.

Then, each complaint will need to be reinterpreted to determine what was meant in the text description and then assigned an index.

(continued)

Interpreting reported complaints

Reported Complaint	Interpretation	Code
The needle is contaminated	Contamination in needle	P1.0 D2.0 T?
There was a leak in use	Leak in product	P? D1.0 T4.0
There was contamination in the fluid path found before use	Contamination in product	P1.0 D2.0 T2.0
The cannula leaked on setup	Leak in cannula	P1.0 D1.0 T3.0
The tube was separated	Leak in joints	P1.0 D1.0 T?

The table provides an example of how a set of textual data must be interpreted and then coded according to the index scheme in Table 9-1. In many entries, critical information is missing on either the part or time. When analyzing existing textual data some indices will be left blank.

Once the existing data is mapped and the indexing scheme updated to reflect any lessons learned, the complaints tracking can be switched to the new system. The examples from the mapping process can be used in training materials.

- **By Date of Occurrence.** The date when the failure happened may give some indication of which lots have a higher defect rate. However, a delay between the occurrence and when it is reported may skew the apparent short-term failure rates. If complaints tend to trickle in over a long period of time, graphs may be skewed and show improvement where there is none (i.e., the complaints happened, you just haven't heard about them). In order to account for this, the defect rate should be adjusted based on the expected delay in reporting.
- **By Date Produced.** Data tracked this way may show trends in defects rates with time. However, products may sit in different customers' inventories for widely varying times time before being used, and the time delay will impact the ability to identify trends.

9.9.3. Review Quality Control Data

A single group should be responsible for monitoring any quality data that comes out of the production environment, including that from SPC, tests, inspections, and scrap and rework. Measurement plans are often set up and carried out without assigning one group the responsibility of monitoring the data and ensuring that process capabilities are acceptable and improving.

9.9.4. Track the Impact of Changes

Suppliers, manufacturing processes, materials, and part designs will change during the life of a product. As cost is reduced through efforts other than variation risk management, changes may have unexpected effects on product quality. When any changes in the product or its manufacturing methods occur, the variation flowdown should be reanalyzed to determine what the effects on the product's cost and yield will be. Ideally, no change should be permitted until a VRM review approves it.

9.10. SUMMARY

In this chapter the VRM methodology for each stage in product development is reviewed. At each stage, the team should work to create the optimal design and process that satisfies both the target requirements and the target tolerances for the KCs. Using variation flowdown methods and assessment tools, teams can quickly identify the most critical product, system, part, and process KCs. Then, appropriate and cost-effective mitigation strategies can be chosen and implemented to optimize product yields and costs.

10

ROLES AND RESPONSIBILITIES IN VARIATION RISK MANAGEMENT

Successful implementation of variation risk management is highly dependent on the people involved and on the company's organizational structure. There are a number of ways of implementing VRM. Each company should implement the methodology that best suits its culture, its existing product development system, and its team structures. As will be discussed in this chapter, every function within the company has a responsibility to identify and remove sources of variation and to reduce the impacts of variation. The IPT combines the cross-functional expertise and tool kits of its members to most effectively design and build new products or improve existing ones.

If just one functional group is involved at a given stage of product development, the solution chosen will likely be optimal for only that functional group. It takes all groups working together to determine the best solution for the company. The team must be structured to bridge the gaps during the product's development and production. Each member of the team has the responsibility to determine where his or her function can contribute to the solution and to identify how the proposed solution may impact his or her function.

Roles and responsibilities of the various team members are different for new products under development than for products already in production. During new product development, the team members bring expertise from a variety of disciplines and experience to identify *potential* sources of variation and their likely impacts on product quality and cost. During production of an existing product, the team members bring information about the *existing* costs and impacts of variation on costs to their groups as well as identify the potential impacts of any proposed design or manufacturing process changes. In both cases, involving multiple functional groups increases the chance that the optimal solution will be found. When using teams, it is important to pay attention to the organizational and team dynamics (Cleland, 1996; Fisher, 1999; Ulrich and Eppinger, 1995). The last section in this chapter covers roles and responsibilities for different types of suppliers.

10.1. PRODUCT DEVELOPMENT

The application of variation risk management during development of a new product requires a multidisciplinary and cross functional approach. The identification, assessment, and mitigation procedures all require designers, manufacturing engineers, and production personnel to work together. A frequently used approach is an integrated product team; however, some companies have successfully created specific teams of experts whose only task is to manage and address variation-specific issues while functional teams focus on the product development or improvement efforts. Organizations may also opt to use a coach-based approach. The next three sections describe the benefits and drawbacks of each approach.

10.1.1. The Integrated Product Team Approach

The integrated product team approach uses the existing IPTs to execute the VRM methodology. The IPT members should work together to identify, assess, and mitigate the risks that variations will negatively impact the final product cost and quality.

There are benefits and drawbacks to using IPTs. The benefits include:

- The entire team understands and participates in variation risk management and can apply the tools on a day-to-day basis.
- The entire team agrees on the critical problems and solutions. As the team goes forward with the design, the members agree on prioritization and approaches. The team can avoid confusion, changing priorities, and finger-pointing at later stages.
- Ultimately the VRM methodology should reduce the time spent on quality improvements by focusing the IPT on the most critical issues. In addition, it should reduce the constantly changing focus due to the "problem of the week" effect.

Drawbacks include:

- The entire team will require training and the associated time investment. In addition, the quality of the VRM methodology will differ from one

IPT to another: Some teams will be better and/or more enthusiastic than others.

• The identification, assessment, and mitigation procedures can be time consuming. For each of the KCs, only a subset of the team typically is needed. However, to ensure that variation risk management works, the entire team needs to be available for all discussions, and participate in periodic meetings.

A typical IPT consists of members representing the various functional groups within a company or division (Fig. 10-1). The members can include design engineers from a variety of disciplines as well as people from manufacturing, quality, systems engineering, and suppliers. The IPT leader has ultimate responsibility for the system delivery. A product manager typically will be responsible for a set of IPTs and the finished product. The systems engineering function has the responsibility of ensuring that individual systems can be integrated together.

The IPT should (1) design the product and its manufacturing processes concurrently and (2) include all relevant functional groups early in product development. Using the preliminary information released by the design groups, the manufacturing organization can get a head start on developing the production environment. The team can make choices resulting in better manufacturability, producibility, and cost before the design is locked in. The net result will be a shorter product development cycle.

Following are the roles and responsibilities of the various participants in product development. (See also Table 10-1.)

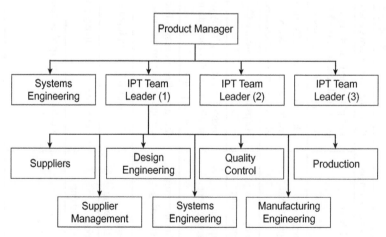

Figure 10-1. Integrated product team structure.

Table 10-1. Roles and responsibilities in product development

	Requirements Development Concept Design	Product Architecture Design System Concept Design	Detailed Design Testing and Refinement
Product manager	Runs the gate reviews and ensures that the VRM methodology is completed. Provide incentives for applying VRM.		
IPT lead	Facilitates and encourages the team to create the flowdown. Appoints a person responsible for maintaining documentation. Checks the completeness of the KC documentation and verifies that all potential problems have mitigation plans. Prepares and presents materials for gate reviews.		
Systems engineering	Identifies product KCs, flows down requirements to systems. Identifies critical interfaces.	Ensures the flowdown and mitigation plans are consistent with the product requirements. Provides feedback on the relative importance of system KCs. Provides feedback on integration and interface issues.	
Design engineering	Helps identify system KCs. Creates the flowdown and assessment for systems. Identifies design changes to improve robustness.		
Supplier management	Identifies suppliers likely to impact quality. Identifies the data needed from suppliers.	Works with suppliers to ensure robust designs. Gathers supplier capability to support assessment.	
Manufacturing engineering	Identifies key processes likely to have a major impact on the system KCs.	Collects data to support the assessment activities.	Validates high-risk processes. Provides data for assessment procedures. Implements improvements in manufacturing.
Tooling engineering	Uses part KCs as an input to tooling design. Provides tooling capability data.		Ensures the tooling reliably delivers the part and process KCs.
Process development	Communicates the likely process capability to the design team for new processes.		Ensures the new processes reliably deliver the part and process KCs.
Quality control	Identifies where quality control may be a potential mitigation strategy.		Ensures the quality control plan is consistent with high-risk KCs. Creates a validation and mitigation strategy for production.
Suppliers	Identify high-risk part KCs. Make recommendations to teams for improvement.		Ensure the delivered parts or systems will meet quality targets.

- **Product Manager.** The product manager's role is to oversee the entire product development process. He or she should actively review the results of the variation risk management process in the gate reviews (Chap. 9). The manager should also provide incentives for proper application of variation risk management and should reward IPTs for proactively addressing variation.
- **IPT Leaders.** The IPT leaders have responsibility for ensuring that the impacts of variation are managed throughout product development. Each leader should work with his or her team to conduct the I-A-M procedures at each stage.
- **Systems Engineering.** The role of systems engineering is especially important when a product is divided into many systems. They are responsible for flowing product KCs to the individual system IPTs and identifying variation senstive interfaces.
- **Design Engineers.** The design engineers[1] (e.g., mechanical and electrical engineers) in the IPT have most of the responsibility in the early phases of product development. They are responsible for creating variation flowdowns and working with manufacturing to assess the relative risks of not achieving the target specifications and tolerances. In addition, they are responsible for identifying and implementing design changes to minimize the negative impacts of variation or variation in system KCs.
- **Supplier Management.** Supplier management (also called materials or procurement) should bring the voice and experience of the suppliers to meetings when the suppliers themselves cannot attend. In many cases, parts and assemblies are not built in house but supplied by outside vendors. Supplier quality can have a significant impact on final product quality; however, it is impossible to ensure that every feature of every supplied part can be monitored and managed. The procedure of identifying key characteristics can help focus the suppliers on the most important features. In addition, supplier management is responsible for gathering information about the supplier's capabilities for inclusion in the assessment phase. Supplier management is also responsible for VRM training of critical suppliers and ensuring that VRM is applied at their facilities.
- **Manufacturing Engineering.** Manufacturing engineering should be heavily involved throughout the VRM procedure. Manufacturing engineering

[1]The term *design engineers* is used in a variety of ways in industry. In this book we mean the people responsible for developing the design concepts, not the people who create the CAD models.

members should be included early to enable them to understand which key manufacturing processes are likely to have major impacts on the system requirements. Also, manufacturing groups often have experience with previous generations of the same product and can provide a historical perspective on typical defects, their causes and impacts. As a design becomes more detailed, manufacturing should collect data to support the assessment activities. During the later phases of product development, the manufacturing engineers should validate the high-risk processes and begin to implement manufacturing improvements.

- **Process Development.** For some products, the organization may require the development of a new manufacturing process or the significant modification of an existing one. Process development, also called *advanced manufacturing,* has the responsibility for communicating the potential process capability and ensuring that the manufacturing process is capable of delivering the required quality.

- **Tooling Engineering.** Often there is a group specifically responsible for designing the product-specific tooling (e.g., dies, fixtures, and molds). Because the design and fabrication of tooling often require long lead times, tooling design must start early in product development. The variation flowdown process will identify the critical part and process KCs that are driven by tooling. The tooling designers, like the product designers, should work to achieve the target requirements as well as target tolerances for these critical KCs. If the KCs are identified early, the tooling design is more likely to deliver the part quality required by the product.

- **Quality Control.** Quality control may not have direct input in the early phases. However, like manufacturing, it can develop an understanding of what and where quality control may be needed during production. As the detailed product design is completed, quality engineering is responsible for developing a quality control plan to mitigate the high-risk KCs.

- **Suppliers.** Critical suppliers should be included in the VRM methodology (see Sec. 10.3). The suppliers can make recommendations early in product development to reduce the source and impact of variation. In the later stages, the suppliers are responsible for ensuring that the parts or processes they provide are of the appropriate quality.

- **Other Functional Groups.** A variety of other functional groups can play a role in VRM. Finance can provide information about the cost of various mitigation options. Production can provide information about production costs and the impact of mitigation strategies on capacity. Prototype fabrication groups provide information about the preliminary product performance. In addition, the IPTs may call on people with specific capabilities, such as statistical experts, variation risk management experts, and consultants, to help throughout the I-A-M procedure.

In summary, all functional groups should contribute two key inputs. The first is identifying sources and impacts of variation. The second is working within their functional responsibilities to reduce sources and impacts of variation. The inputs of all functional groups are needed to best develop the holistic view of variation and the optimal design that simultaneously minimizes cost and maximizes quality and performance.

10.1.2. Expert Teams

In the IPT method, the functional groups must meet to discuss the sources, impacts, and trade-offs relating to variation. Because the VRM methodology can be perceived as time consuming, some companies or divisions have opted to take the approach in which the methodology is applied by a small team of experts whose specialty is variation risk management.

These are often called tiger teams (Wheelwright and Clark, 1992) or Six Sigma Black Belt teams. These small teams work with individuals in or subgroups of the IPTs to develop variation flowdowns, assess the risks, and help the IPT reduce risk. Even though the IPTs provide information, the analysis is owned and maintained by the expert team. Variation risk management experts will work with the entire IPT for short periods of time to present results and get feedback.

There are benefits and drawbacks to the expert team approach. The benefits are:

- Variation risk management experts can develop variation flowdowns and assessments more quickly because of their previous experience.
- The time requirements on the IPT are significantly less.
- The quality and completeness of work is higher and more consistent across systems.

However, these benefits do not come without problems:

- The IPT may not accept the VRM methodology. Although the experts may present the critical issues to teams, there is a risk the IPTs will not agree with, or follow up on, the recommendations.
- The IPTs are not well educated in the VRM methodology and may not apply the tools in their day-to-day decisions.

10.1.3. Coaches

Another approach is to create a group of variation risk management coaches. In this approach, each entire IPT is responsible for variation risk management procedures; however, the members do not go through the entire VRM procedures as a team. Rather, teams are given a small amount of training and work is divided among the team members. The coaches act as the facilitators in the IPT meetings and work with individuals to carry out the I-A-M procedures. The benefit of this approach is that the impact on the team is lower because the training can be shortened and teams have a shorter learning curve. The drawback is that the IPT may not gain the full benefit of having done the VRM methodology itself. Coaches ensure a more consistent output from the various functional teams. However, the coach, like the expert team, does not have the ability to ensure the implementation of suggested mitigation plans.

Management should reward those who successfully coach teams. Often, good people who are put in the role of coach do significant work to help improve the design but do not get rewarded for their efforts. It is like the situation of a coach for any sports team—if the team loses, it is the coach's fault. If the team wins, the team members did it.

10.2. PRODUCTION

The roles and responsibilities in the production environment are not as complex as during product development. The functional groups involved in the VRM process may include:

- **Production Management.** These are the people responsible for managing the day-to-day activities on the production floor. They will typically understand the production flow and where QC, scrap, and rework occur.
- **Quality Control.** These are the people responsible for the current quality control of the product. They own quality plan and the data relating to the current quality, yields, measurements, quality control plans, and so on.
- **Finance.** The financial people own the data required to determine the total materials used and total numbers of good products made and shipped. In addition, they will provide the data relating to costs and costing methods.
- **Engineering/Design.** Often members of the engineering or design staff are assigned to existing products. Their responsibility is to implement any design changes or improvements and to provide engineering sup-

port for any rework or scrap decisions. They are often the best source of information about design intent of the product.

- **Plant Manager.** The plant manager has a vested interest in reducing the cost of manufacturing the product. He or she should be involved in identifying and tracking critical improvement projects. The plant manager is the strongest driver and should ensure that proper implementation of VRM is rewarded.
- **Manufacturing Engineering.** Often there is a team with specialized expertise in specific manufacturing technologies. These people have the expertise to do the manufacturing process improvements needed to reduce variation introduced by the production equipment.

The culture of the organization will greatly influence the relative difficulty of retrieving the relevant data from the production floor. Groups may be unwilling to share critical information that may make them look bad. The team should avoid the temptation to assign blame and focus on future improvements.

As with VRM in product development, there are two approaches to VRM in production—using VRM-trained production teams or VRM experts.

10.2.1. Production Teams

The production teams are groups of people drawn from the preceding list who, as a team, perform the identification, assessment, and mitigation activities. The team is responsible for the development and execution of any plans. The benefit of the team approach is that information and analysis are likely to be complete and to include information from all functional groups as well as the field. However, the success of the projects will be highly dependent on the skills of the team members and their attention to the VRM process. In addition, there is a fairly steep learning curve to efficiently and correctly integrate the various data sources. While training can help facilitate learning VRM, bringing the average skill base up can be expensive and time consuming.

10.2.2. Expert Teams

A VRM expert team, also called a tiger team (Wheelwright and Clark, 1992) or Six Sigma Black Belt team (Pande et al., 2002a), typically works with a production team on one or two projects at a time. The benefit to this approach is that a project is done more quickly and with a higher degree of consistency. However, because the people responsible for production are only peripherally involved, the follow-up efforts to maintain the improvements may not be sufficient to maintain long term benefits.

10.3. SUPPLIERS' ROLES AND RESPONSIBILITIES

In addition to the functional groups within the organization, suppliers play a critical role in VRM. It is not unusual for suppliers to provide at least 70 percent of the components of finished products. Because suppliers can have a large impact on final product quality, they must be treated as integral members of the IPTs. A supplier's ability to deliver high-quality parts may be more important than the in-house groups' abilities to control variation. When there is a problem with suppliers' parts, significant overhead is required for incoming inspection, meetings, root cause analyses, contract renegotiations, and so on.

Traditionally, the lowest bidders were chosen as suppliers. Today, that is changing for many organizations as they began to understand the hidden costs associated with using price as the primary selection criterion. First, product development organizations are requiring their suppliers to have demonstrated their ability to control the quality of their product through such processes as ISO-9000. Second, product development organizations are choosing the suppliers that have a proven track record in their ability to deliver consistent quality.

Most U.S. and overseas firms have comprehensive supplier management processes. For example, Boeing Commercial Aircraft uses their Advanced Quality System (AQS). These processes typically outline the responsibilities and requirements for suppliers and have the goal of ensuring the delivery of high-quality parts and systems. Most supplier management processes define how the suppliers prove process capability during transition to production and during production. Quality control reports from the supplier are usually used to ensure compliance with specifications. These reports can include SPC data, statistical sampling, and/or inspection and test results. Some also outline how suppliers should operate in product development; however, requirements for supplier product development processes tend to be general and are typically left to the discretion of the supplier.

It is not necessary or feasible to include all suppliers in the VRM methodology. Only those whose parts or systems directly and significantly contribute to variation in system KCs should be involved. The suppliers involved in VRM during product development or during production should have basic training and documentation to educate them on the following:

- **The Impact of Variation in Their Processes on the Final Product.** This education can include reviewing the flowdown, cost impacts, and risk assessments done as part of the I-A-M procedures.
- **The VRM Methodology.** In order to improve the system or part being delivered, the supplier must know and apply same VRM tools being used by their customer.

10.3.1. Role of Suppliers during Product Development

This section outlines how suppliers can be integrated into product development and the VRM methodology. The interactions with each supplier will be highly dependent on two factors: their contributions to the high-risk system and product KCs, and the type of supplier.

Relative Contributions to Variation

It is tempting to treat all suppliers the same, but, given the amount of time it takes to set up meetings, review processes, and so on, it is not economically feasible to fully integrate every supplier into the VRM methodology. The output of assessment during the early stages of product development should be used to identify which suppliers to integrate and which issues can be handled off line by supplier management. The choices will be based on (1) the relative risks associated with system KCs and (2) the relative contributions of the suppliers to the total variation (see Chap. 6).

Type of Supplier

There are different types of suppliers. Some will provide commodity parts and materials, and others will be responsible for the design and production of specialty parts or major systems. While there are many ways of dividing up the world of suppliers, the following are typical of the range of relationships.

- **Commodity Suppliers.** These suppliers sell parts off the shelf. There is little room to change their designs or production processes. Parts or assemblies are picked from catalogs or assemblies are made from standard components.
- **White-Box Suppliers.** These are build-to-print suppliers who take designs developed by the IPT and build them. For example, printed circuit boards and stamped sheet metal for cars are often designed in house and then manufactured by an external supplier to save on capital costs and to leverage supplier expertise in specific manufacturing processes.
- **Black-Box Suppliers.** These suppliers are provided with the interface and system requirements and are given complete control over design and production of their parts and systems.
- **Tooling Suppliers.** The design and manufacturing of tooling (e.g., stamping dies) may often be outsourced to suppliers. This is done to shorten the lead times and leverage expertise. Typically, a team will provide a drawing of the part to be made and the supplier will design and produce the tooling to deliver that part. Tooling suppliers may provide input on what system features may be easier or less costly to manufacture.

The next sections review the role of each supplier type in VRM and details the specific tasks for each of the identification, assessment, and mitigation procedures.

Commodity Suppliers

Commodities can include such items as standard motors, chips, screws, or connectors. There are typically many suppliers that can deliver the same part with minor differences in geometry, expected quality, and requirements. Commodity suppliers are also often capable of delivering customized parts or standard parts with customized tolerances. When purchasing products where different quality levels are available, the team should ask how the higher-quality products are produced. If the higher-quality products are produced using more precise manufacturing processes, the distributions are likely to be normal Gaussian distributions. If higher-quality products are selected from a single pool of products, the distributions may not be Gaussian (see Sec. 8.2.5). The latter scenario can cause unexpected degradation in quality because more parts are close to the tolerance limits than would occur if the distribution was Gaussian. Commodity suppliers will typically be brought in at the later stages of product development.

Identification. During identification, parts supplied by commodity suppliers will be the last elements on the ends of some of the branches of the variation flowdown. These part KCs should be communicated to the supplier. The information should include (1) the criticality of the feature and (2) the acceptable variation in the form of a tolerance and a C_{pk}.

Assessment. Assessment can be based on three data sources. First, procurement and production team members may have information about the quality of similar parts used in the past. Second, if incoming inspection was used on previous designs, this data can also be referenced. Third, if no in-house data is available, suppliers will often supply data sheets about the specifications of parts. These data sheets often include the expected variation in the delivered parts. If this data is available, the team should look into the shape of the distribution as well as the percentage of defective parts. This is important if the specification data is being used to populate variation models.

Mitigation. Little can be done to fix a commodity part or assembly: In most cases, what you buy is what you get. The team can decide to either purchase more expensive parts or selectively choose high-quality parts from a large population. It also may be necessary to verify that parts delivered comply with the specifications promised by the supplier.

White-Box Suppliers

In the case of white-box suppliers, the product development organization maintains control of the product design but uses the manufacturing capabilities of the suppliers to deliver parts or systems. The suppliers may be responsible for developing the manufacturing equipment and/or tooling. Typical examples of products purchased from white-box suppliers include printed circuit boards, injection-molded parts, and stampings. The white-box suppliers have significant expertise in the manufacturing of products and can provide input on the manufacturability of parts and the ability to achieve the target requirements.

Identification. Because the IPT owns the design of the part or assembly, the in-house team can complete the entire variation flowdown to the part KCs. However, the supplier should have input into the process. This is important because the supplier may be aware of interactions or additional sources of variation that are not apparent to the design group. The output of the identification phase will be a list of part KCs for which the supplier needs to ensure process capability.

Assessment. Assessment will require the involvement of the suppliers. They have firsthand knowledge of their manufacturing processes, and in the early in product development they can provide input on the design of individual parts or assemblies to reduce the impact or magnitude of the expected variation. For example, the supplier may give recommendations for different chip configurations on a printed circuit board that may improve the yield. Other sources of assessment information may come from a history with the supplier, industry standards, and supplier-provided process capability data.

Mitigation. All tools of mitigation are available in the case of a white-box supplier. The in-house team can change the design to better match the manufacturing processes. The supplier's processes can either be changed or improved; however, this type of change will typically increase costs. Using supplier management requirements, the IPT can impose special quality control.

Black-Box Suppliers

Using black-box suppliers leverages design and manufacturing expertise that may not exist in house. In the case of critical assemblies, black-box suppliers should be included as active members of the IPT. Involving the suppliers in training and helping them apply the VRM methodology better ensures final quality, cost, and performance. The I-A-M procedures used by the suppliers will be the identical as those used by the internal teams (Chapter 9).

Tooling Suppliers

The tooling supplier provides design-specific tooling for everything from printed circuit board tooling to stamping dies. Typically, the customer provides the design of a part to the tooling supplier, who designs and fabricates the tool. The supplier's tooling designers can provide valuable input as to the manufacturability of the design.

Identification. The design of the part and the part KCs identified during the flowdown process should be communicated to the tooling supplier as early in product development as is practical. The target values of the KCs and the permissible distributions of values about the means (or upper and lower limits) should be clearly delineated to the supplier.

Assessment. Assessment will require the involvement of the tooling suppliers. Based on the design and the part KCs, the supplier can provide input as to the achievability of the part KC tolerances. This step may include prototype tooling or process capability analyses based on previous designs.

Mitigation. The tooling supplier can provide suggestions on how the design can be changed to reduce variation introduced by the production process. Iterations in part design or modifications to the tooling may be needed to meet part KC tolerances.

10.3.2. Role of Suppliers during Production

It is easy to blame suppliers for all problems with quality. However, proving it can require significant data and analysis. Assessment can be used to demonstrate sources and causes of ongoing quality problems and excess cost. If a problem is discovered, the supplier management processes described earlier can be used.

A number of methods can be used to control variation in parts delivered by the supplier:

- **Deploying VRM Teams to Help the Supplier.** The best method is to help the supplier improve its processes. Often suppliers do not have the in-house expertise to improve their own processes. Sharing VRM experts with the supplier and working closely to help improve the supplier's processes can lead not only to the desired part quality but also to reduced costs, and improved relationships.
- **Contractual Obligations.** The expected variation and acceptable quality should be built into the contract language. Cost savings from continual improvement can be shared between the supplier and customer to encourage VRM efforts.
- **Incoming Testing and Sampling.** Depending on the maturity of the supplier, the history with the supplier, and the criticality of the part, the incoming material can be evaluated. While inspection and testing is expensive and should not be used as a matter of course, in some cases it is cost effective. It can also generate good data about incoming quality that can be compared to the data collected by the supplier. Reconciling the

differences between in-house and supplier data will ensure that both are using similar measurement systems and that a minimum number of defective parts will leave the supplier.

- **Excess Inventory.** If the quality of the supplier's parts is not stable, it may be necessary to carry extra inventory to ensure that production lines do not become starved.
- **Buying More Precision than You Need.** By purchasing higher-quality parts than you need, you can buffer against any degradation in quality.

All but the first method add recurring cost and overhead and should only be used if absolutely necessary.

10.3.3. What Does a KC Mean to a Supplier?

When working with a first-tier supplier, one manufacturing engineer drew Fig. 10-2 and asked, "What do I need to do differently?" The supplier was being required to reduce variation in the key characteristics, but already had an active statistical control program. The engineer's point was that the supplier would, as part of its quality initiative, ensure that its processes were in control and that all parts delivered were within the allowable tolerance. The engineer's valid concern was the overhead of another process that only increased the supplier's costs and generated paperwork.

The first assumption that everyone needs to challenge is that within tolerance equals good quality. The goal of a KC is to ensure that the delivered feature is on target and has low variation. According to the definition of a KC, the loss function associated with any deviation from target increases cost of the product. There are thousands of dimensions and features that will have tolerances on them, and the supplier will not be able to actively track and control all the tolerances. One purpose of the VRM process is to identify the critical few KCs that should be paid attention to by the supplier. As the supplier improves its processes, it should focus on reducing variation in the KCs of the product first. The KC process should be communicated as a tool to help suppliers focus on what is critical, not as an additional reporting requirement.

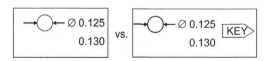

Figure 10-2. How does the addition of a "KC" impact the way suppliers produce the part?

10.4. SUMMARY

Ideally, the IPTs and suppliers will work together to simultaneously improve performance and quality while reducing cost. However, this ideal can be difficult to achieve. Often, because the design group is driving the VRM methodology in the early stages of product development, performance is addressed first, then efforts are made to minimize cost, and finally plans are made to improve quality. VRM and overall cost issues are not properly addressed by this sequential approach, and expensive iterations back to the design stages may be needed. There is often a head in the sand approach that assumes that if the designers just specify tolerances, the production and quality teams will be able to produce to those specifications. Skills and inputs from all functional groups as well as suppliers are required to identify the weakness in a design and determine the most effective solution.

11

PLANNING AND IMPLEMENTING A VARIATION RISK MANAGEMENT PROGRAM

The first eight chapters of this book discuss variation risk management tools and methods. In addition, Chaps. 9 and 10 discuss how VRM can be integrated with the existing product development process and how the functional groups should support variation risk management. Given all of this material, the next question is: How do we do this within our organization? This chapter outlines how to plan and implement an organization-wide VRM program. It is assumed throughout the next few chapters that the VRM program will be implemented in an integrated product team environment.

This chapter is divided into two sections: planning a variation risk management program and its initial implementation. The primary readers for this chapter will be those responsible for implementing organization-wide VRM programs.

11.1. PLANNING A VRM PROGRAM

When faced with implementing a new program, most teams' first impulse is to dive in and start applying tools rather than to do the homework to plan and define the program first. A good analogy to planning a VRM program is painting a room. It is tempting to start right in rolling on the paint, forgoing the dropcloth and masking tape in the assumption that you won't need to tape and will not spill any paint on the floor. Inevitably, when the prep work is not done, you spend more time cleaning up and touching up than if you had properly laid out dropcloths and masked and prepared the wall ahead of time. If teams take time to plan the VRM methodology and do not dive in too quickly, the result will be a more streamlined, useful program that will not require a lot of tweaking and modification later. This section covers the basic steps involved in planning for a VRM methodology. The basic steps are:

- Getting management support
- Gathering organizational support
- Baselining the existing VRM procedures and tools
- Formalizing the VRM methodology
- Developing KC tracking methods
- Identifying lead users
- Developing training materials

11.1.1. Gathering Management Support

Small groups within an organization are often tempted to develop and deploy a variation risk management program without adequate support from management. If management approves VRM efforts, but does not make firm commitments to provide resources, promotions (if successful), or incentives to the rest of the organization to adopt the methods, VRM will probably fail. Without strong management support, many hours and resources can be spent developing VRM methods only to find that the rest of the organization is unwilling or unable to adopt them.

Management support is a catchall phrase used to describe a variety of organizational behaviors and methods. In a VRM methodology, management should provide support in a number of ways:

- **Resource Allocation.** The implementation of a VRM methodology requires money and staffing. Funding should be identified and earmarked before planning the development and implementation of VRM.
- **Promotions and Incentives.** Management teams should reward successful implementation of a variation risk management program with promotions, bonuses, or pay raises. This will encourage the best and brightest to participate and help the program succeed.
- **Facilitating Trade-offs between Functional Groups.** In many cases, the decisions that are good for the organization will benefit the participating functional groups unequally. By understanding and facilitating these trade-offs, management can reassure individual groups, such as the design group, that they will not be penalized for spending extra time or resources if it creates a net benefit for the entire organization, and that they will share in the rewards.

- **Participating in Gate Reviews.** Whether during product design or production, it is imperative the management team ask about the current variation risks and what is being done to fix them. Organizations must not be afraid to delay the progress or launch of a product that has significant variation issues. The management team must direct the IPT to institute variation risk management by making it a clear criterion in the design gate reviews.
- **Encouraging Initiatives.** The behavior of management has been shown to be a key factor in the success of new initiatives. When we were children, "Do as I say, not as I do" fell short when encouraging us to do the right thing. Parents who ask their children to "do as I do" tend to be more successful in encouraging good behavior. The "do as I say" behavior can manifest itself in fast-tracking pet projects around the I-A-M procedures, letting the schedule take precedence over releasing robust designs, and praising individuals rather than the group for success.

Management support is a two-way street. In addition to securing resources and support, the team promoting the VRM methodology should provide and live up to measures of success. These metrics should be reported and made a condition for continued support of the VRM methodologies. Product and process teams should be required to show the benefits realized by applying the VRM methodology. Section 9.1 describes several possible metrics for VRM.

11.1.2. Gathering Organizational Support

The VRM methodology should be developed with input from all functional groups; allowing objections and issues to be brought up before VRM is deployed. This is not to say that the VRM process should be designed by committee—that is a certain way to never finish implementation. However, concerns should be heard and an attempt should be made to address all functional group issues.

In addition, individual groups may already employ VRM tools. When one group develops the VRM process without referencing existing methods, understanding the organizational processes, or taking the company's culture into consideration, the tools are more likely to fail. Successful adoption and learning of a new tools and procedures is made much easier if it complements already existing practices and involves the whole organization.

11.1.3. Baselining the Existing VRM Processes

A common temptation for companies seeking to implement a VRM methodology is to create a new separate process from scratch. The result is a "method of the month" syndrome. Employees are likely to think, "I know I should implement this, but in a year it will just be replaced with another acronym, so I'll just keep doing my job the way I see fit and play along just enough to avoid causing problems."

Rather than create an entirely new methodology, an organization must identify existing variation reduction and management tools that are working and augment those with other methods and training to make the whole VRM process more methodical, holistic, and robust. In addition, the new VRM process must fit in with the existing product development process and team structures.

Baselining the existing VRM processes involves asking the following questions:

- What is the current product development process and how will the identification, assessment, and mitigation procedures fit within it (see Chap. 9.)?
- What are the current team structures? If IPTs are used throughout the organization, they should be used for implementation. On the other hand, if a company tends to be functionally oriented, using experts to implement the VRM methodology may be more successful.
- What VRM tools and methods are currently being successfully used? Companies should take advantage of existing and useful resources. For example, if a company has a variation simulation expert, it can include him or her explicitly in the assessment phase. Also, a company can look for improvement efforts within the organization that have proven successful and use those methods.

The answer to the last question can be answered by the variation risk management maturity model in Appendix A. The maturity model is made up of 17 practices in product development and 9 practices in production (Thornton et al., 2000). Each of the practices has three levels of maturity. By comparing your organization's current practice against the maturity model, it is possible to identify the major weaknesses that need to be improved and the major strengths that should be maintained and encouraged. The maturity models for new product development design and for production share some basic practices but differ in a number of ways.

The design practices that should be evaluated are:

- **VRM Definitions and Methods.** It is necessary to have consistent definitions for identifying KCs and consistent and documented methods for identification, assessment, and mitigation.

- **VRM Training.** Comprehensive training teaches teams how to apply VRM in product development.
- **Integration of VRM into Product Development.** The VRM methodology should be seamlessly integrated into product development. Gate reviews include criteria for identifying, assessing, and mitigating the highest-risk KCs.
- **System KCs.** System KCs must be correctly identified.
- **Variation Flowdown.** Systematic flowdown processes aid in identifying all part and process KCs and creating a holistic view of variation.
- **Ranking.** System KCs are quantitatively ranked based on both the cost and probability of creating a defect. The contributions of part and process KCs are also ranked.
- **Tolerancing.** Tolerances are consistent with system tolerances and are achievable with current capability.
- **New Technology.** Robustness of technology is evaluated prior to product development. Strategies to verify and control the related KCs are in place.
- **Process Capability.** Historical process capability data is used to verify a design prior to production.
- **Uncertainty Reduction.** Uncertainty about risk is reduced through the use of prototypes and variation models.
- **Mitigation Selection.** Mitigation strategies are selected based on their potential benefits, recurring costs, nonrecurring costs, and strategic impact.
- **Robust Design.** Robust design is used to reduce the impact of variation in the most appropriate areas.
- **QC Planning.** Quality controls are planned in concert with variation risk management.
- **Suppliers.** Key suppliers are integrated into the VRM methodology.
- **IPTs.** IPTs work jointly to identify, assess, and mitigate the impacts of variation.
- **Management Support.** Management supports the VRM methodology through resource allocation, incentives, and active involvement in gate reviews.
- **Documentation.** Information about KCs is documented in a clear and consistent format.

The maturity model practices for VRM in production are the following:

- **VRM Definitions and Methods.** It is necessary to have consistent definitions for identifying KCs and consistent and documented methods for identification, assessment, and mitigation.

- **VRM Process Initiation.** VRM should start at the beginning of production and continue throughout the life of the product.
- **Variation Flowdown.** Systematic flowdown methods are used to identify all part and process KCs.
- **Cost of Variation.** Variation costs are quantified and include costs from multiple sources such as quality control, scrap, rework, returns, and production capacity.
- **Mitigation Selection.** Mitigation strategies are selected based on their potential benefits, recurring costs, nonrecurring costs, and strategic impacts.
- **Quality Control Maturity.** The quality control plan allows for defects to be detected while not overburdening the factory with excess measurements.
- **Supplier.** Critical suppliers are involved in the VRM methodology.
- **Management Support.** Management supports the VRM methodology through resource allocation, creating incentives for its application, and monitoring the progress of projects.
- **IPTs.** IPTs work jointly to identify, assess, and mitigate the impact of variation.

Each item has three levels of maturity (1, 2, or 3). In order to assess the company's current status, each practice should be evaluated and the maturity level determined. In some cases, different divisions will be at different maturity levels; and each division's maturity should be determined and included. This allows for identifying the best practices to share between divisions.

Table 11-1 shows the maturity model for VRM definitions and methods. In this example, division A is at a maturity level 1 and does not use the company practices. The rest of the divisions have a common definition available but do not use it consistently. The current definitions and methods should be

Table 11-1. Maturity levels of VRM definitions and methods

Practice	Maturity Level		
	1	2	3
VRM definitions and methods	Teams develop their own definitions and methods. No consensus or commonality between functional groups.	Common definitions and methods are documented, but teams deviate from standards.	All teams use the same definitions and methods.
Company maturity	Division A	Divisions B, C, and D	None

analyzed to determine why they are not being used. The organization can choose to use the existing methods, create incentives to encourage their use, or redesign the methods to make them more user friendly.

11.1.4. Formalizing VRM

Once baselining is completed, the VRM methodology and the way it is integrated with new product development and/or production should be developed and documented. While this book supplies a starting point for creating such a program, some work will have to go into tailoring the VRM process described here to integrate it with the organization's existing product development, production, and business practices. In the beginning, the implementation team should answer the following questions:

- What team structure will be used? Will VRM be executed by the IPT or by a small team of experts?
- How do product development and production management tasks integrate with the I-A-M procedures? For example:
  —What additional elements will be added to the project approval process and/or the stage gate processes?
  —How will the standard design, manufacturing, and quality documents need to be modified to include the KCs and the mitigation plans?

11.1.5. Developing KC Tracking Methods

During the VRM process, it will be necessary to record all information about the KCs, variation flowdown, assessment, and mitigation. Companies often develop their own software or tools for managing and tracking product KCs. Requirements management software packages are commercially available and can be tailored for VRM purpose; however, they have many limitations. They do not allow for complex trees of KCs to be created easily, nor do they easily accommodate the additional information described later.

Typically, KC information is stored in customized Excel or Access database formats. These tools allow for rapid development of the tracking software and easy reporting. When deciding on how to document the information, the following should be determined:

- Common numbering scheme
- Method for recording the variation flowdown along with the assessment and mitigation results
- Format and content of the summary document

Additional information specific to a company's practices and requirements can be added.

Numbering Scheme

A common numbering scheme should be used to ensure that the KCs can be tracked. The numbering scheme should be determined before the start of VRM. Figure 11-1 shows one method for labeling KCs uniquely. The first three letters (AAA) represent a unique code for the IPT's system. The second three letters (BBB) are used to indicate the assembly or part to which the KC belongs. The three numbers (###) uniquely identify the KC.

Some teams will be tempted to group KCs into categories such as safety, manufacturing, and engineering. While using categories can help sort KCs into groups, teams can waste significant time arguing over the categories rather than working to improve the product. In addition, KCs often fall into more than one category. The relative importance of the KCs should be handled through assessment.

VRM Documentation

The system KCs and variation flowdown should be documented. A number of different methods can be used to record the flowdown; one is a graphical representation (Fig. 11-2). Complex flowdowns can be large and the graphical representation can be difficult to manage. However, it is much easier to use the graphical representation to facilitate discussions about variation.

Another way to document the variation flowdown is in a table that links the KC and its parents (note that a KC may have multiple parents if it contributes to more than one system KC). Tables 11-2 through Table 11-5 show several entries describing the relationships in the flowdown, the assessment and mitigation documentation.

For system KCs, the target values and allowable variation should be recorded and the level of risk documented. For each assembly, part, and process KC, the relative contribution to the system KC risk should be recorded (Tables 11-2 and 11-3).

For each KC subjected to a mitigation plan, the following should be documented (Table 11-5).

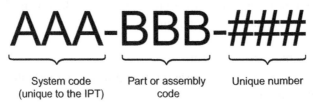

AAA-BBB-###

System code   Part or assembly   Unique number
(unique to the IPT)   code

Figure 11-1. Common numbering scheme.

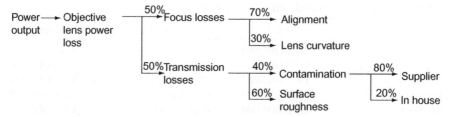

Figure 11-2. Variation flowdown in graphical form.

- The system KC being improved
- The KC being addressed
- The mitigation plan
- The expected completion date
- The person responsible
- The expected reduction in variation

The IPT or project manager should use the summary portion of the KC document to track the progress. The summary should include:

- The number of system KCs identified
- The number of system KCs with complete flowdowns
- The number of system KCs with uncertain risks
- The number of mitigation plans in place
- The number of mitigation plans that have been executed

11.1.6. Identifying Lead Users

Implementing a VRM methodology will require resources and time. To simplify implementation as well as to help introduce VRM to the company, the program should be rolled out to lead users as a trial before being implemented across the entire organization.

Today's products contain a mix of electronics, mechanisms, optics, and structure, all of which can be impacted by variation. It is important to identify one or two functional areas where costs and quality issues are significant. By focusing on these limited areas, it will be possible to develop better-tailored training and examples. It is also important to find a team that is likely to be receptive to changes in its practices. The lessons learned, success stories, and benefits gained can be used to help sell and implement VRM elsewhere in the organization. In addition, a plan for training teams and deploying the methods throughout the organization should be developed.

Table 11-2. Variation flowdown in tabular form

KC	KC ID	Parent Name	Parent ID
Power output	SYS-OPT-001	NA	NA
Object lens power loss	OPT-OBJ-001	Power output	SYS-OPT-001
Focus losses	OPT-OBJ-002	Object lens power loss	OPT-OBJ-001
Transmission losses	OPT-OBJ-003	Object lens power loss	OPT-OBJ-001
Alignment	OPT-OBJ-004	Focus losses	OPT-OBJ-002
Lens shape	OPT-OBJ-005	Focus losses	OPT-OBJ-002
. . . .			

Table 11-3. Assessment/system KCs

KC	KC ID	Cost	Probability of a Defect	Bucket
Power output	SYS-OPT-001	7	5	High
Beam evenness	SYS-OPT-002	9	2	Medium
. . . .				

Table 11-4. Contribution of part and process KCs

KC	KC ID	System KC	Percent Contribution to Parent	Percent Contribution to System KC
Object lens power loss	OPT-OBJ-001	Power output	100%	
Focus losses	OPT-OBJ-002	Power output	50%	
Transmission losses	OPT-OBJ-003	Power output	50%	
Alignment	OPT-OBJ-004	Power output	70%	35%
Lens shape	OPT-OBJ-005	Power output	30%	15%
Supplier particulates	OPT-OBJ-007	Power output	80%	16%
In-house particulates	OPT-OBJ-007	Power output	20%	4%
. . . .				

Table 11-5. Mitigation plans

KC	KC ID	System KC	Type Plan	Description	Responsible
Alignment	OPT-OBJ-004	Power output	Design	Evaluate at different alignment mechanisms	Mechanical engineering
Surface roughness	OPT-OBJ-005	Power output	Process change	Use a higher precision grinding method	Supplier A

11.1.7. Developing Training Materials

Training courses should be developed and used with each IPT. The following are characteristics of a good training program:

- **Efficient.** Although a great deal of depth and detail in a course is desirable, the realities of personnel schedules and limited resources must be taken into consideration. To that end, the training must be designed to transmit the most critical information efficiently. Once the team begins to apply the methods outside of class, trainers and/or coaches should be on hand to answer questions and help apply the VRM process.
- **Just in Time.** The effectiveness of training drops as the time between training and its application increases. Ideally, the training should start just before teams need to start applying VRM.
- **Relevant.** Relating training to an individual's experience and knowledge increases its effectiveness; training that uses examples from outside a trainee's field is often less effective. One reason for focusing initial projects on one or two functional teams is that examples can be generated from the initial deployment and used with other teams.
- **Interactive.** It is well known that chalk talk training is highly ineffective. Trainees need to interact with the material, apply it immediately, and get immediate feedback to learn the material well. To ensure that the people taking the class can both understand and apply the tools being taught, the training should allow the students to practice the materials on their current day-to-day tasks.
- **Trained by Engineers.** The people doing the training should have a technical background. Because the training typically is interactive, the trainer needs to be able to understand the technology in order to facilitate the discussions.

Training can be structured in one of two ways: generalized and just-in-time. In the first approach, everyone in an organization or team receives generalized training. In the second approach, teams receive just-in-time training as they prepare to apply the tools.

Generalized Training
Generalized training is given an entire organization simultaneously. The training can be deployed to people regardless of what their current responsibilities are or the current state of their projects. The benefits of generalized training are:

- It is easier to schedule training around personnel schedules. People can choose classes that fit their schedules from among the many classes offered.

- It is easier to schedule trainers. Trainers can teach the same course several times in a short period. The scheduling can be arranged to accommodate other training requirements.

The problems with generalized training are:

- People forget material in the time between learning it and having to apply it.
- The examples or exercises used in class might have no direct relevance to the work that trainees are doing at the time or to their upcoming projects. Also, if people from different functional groups attend the same class, more examples are needed to illustrate the application of the tools and methods.

Just-in-Time Training

Just-in-time (JIT) training provides tailored training to IPTs as a group just prior to their need to use it. While JIT training is harder to schedule for those taking the classes and those teaching them, the JIT approach carries many benefits:

- People learn the material as they need it and there is less time to forget the subject matter.
- Examples can be drawn from ongoing projects to illustrate VRM applications and methods.
- A team can "do its homework in class." Trainers can also act as facilitators to guide teams through the identification, assessment, and mitigation phases on their current projects.[1]

Because the trainer is working as a facilitator, it will be harder to use people who specialize in training only. Typically, the course should be taught by someone with engineering experience who can help a team work through issues related to its current projects and problems.

11.2. IMPLEMENTING THE VRM PROGRAM

Once the VRM methodology is developed and documented and the training is developed, it must be tested and honed. Implementation should be initiated

---

[1]Another advantage to the just-in-time and doing-homework-in-class approaches is that the time in training can be billed to the project rather than overhead.

in a rollout phase, in which a limited number of teams are taught the VRM process. Based on the teams' feedback, the training and VRM process can be modified and refined.

11.2.1. Identifying Initial Projects

The VRM methodology should be applied to systems that are small enough to be manageable but large enough to present significant opportunities for improvement. For projects related to a new product, a typical product or subsystem will be developed by 4 to 10 engineers. For projects related to production, each team should evaluate only one functional area, manufacturing cell, or assembly. In both cases the project should present a significant opportunity for improvement and the team should be receptive to learning and applying new skills.

11.2.2. Training the Team

Once the projects are chosen, system IPTs or production IPTs should meet together for training. The team should include key suppliers. Depending on the role and availability of the suppliers, their interests may be represented by the supplier management group. However, if a supplier is responsible for critical design and technology, including them in the team would be beneficial. The implementation team should ensure that all members of the IPT attend the training. Before the training, the IPT lead should assign team members the following responsibilities:

- **VRM Owner.** This person has responsibility for following up to gather any information related to KCs that could not be collected in the training. This person also should be responsible for preparing any presentation materials for reviews and for reporting the status of the VRM effort to the team at each meeting. They should arrange follow up meetings to complete the VRM process.
- **Recorder.** This person will be responsible for recording all information relating to the I-A-M process and maintaining it in a location accessible by all team members. In addition, he or she will be responsible for sending out meeting minutes with reminders and action items, due dates, and team members' responsibilities.

11.2.3. Applying VRM

By following the I-A-M steps outlined in the rest of the book, each team should work to reduce the sources and impact of variation on the cost and quality for its project. Detailed records of the variation flowdown, assessments, and mitigation plans should be kept because these records will be used to provide management with feedback on the relative success of the projects. The VRM implementation team should follow up with each IPT, continue to support the process, and help management with reviews.

11.2.4. Gathering Feedback

The first teams to implement VRM should critique the current process and providing feedback on its strengths and weaknesses. Based on the feedback from the team and its trainer/facilitator, the VRM process should be modified and improved to make it more usable and effective.

11.3. SUMMARY

When planning and implementing a VRM program, it is tempting to start defining the process without evaluating what the current tools and methods in use are. Including the best process and identifying the weaknesses in the current tools encourages organizations to accept the tools.

Implementing a VRM program is a three-step process. First, the organization needs to do significant planning to ensure a successful rollout; second, the VRM process should be rolled out; and finally, feedback from teams should be used to improve the VRM process.

12

SUMMARY

This book outlines a set of tools and methods that will allow any team to more effectively apply existing variation reduction and robust design tools. The ultimate goal of variation risk management is to improve product quality, operational efficiency and productivity. This book focuses on how to get more out of your current variation reduction and quality improvement efforts and how to design products of high quality that can be produced reliably. The emphasis throughout is on the efficient and effective use of limited resources. Decisions on where to focus limited resources should be data-driven and based on current process capability and the total cost impact of variation.

A word of caution—do not let variation risk management program become burden on the organization. As Bardwick (1995) points out in *Danger in the Comfort Zone,*

> Cultures create elaborate programs without serious regard to their outcome, because spending money is perceived as a solution. When the process becomes the point, it is not only ineffective, but also hugely expensive. An elaborate process becomes a pork barrel, a business unto itself, protected by those it feeds.

It is important to design a variation risk management program that is effective yet simple and does not increase the workload for engineers too much. There are several key points that teams should keep in mind as they apply these methods. As pointed out throughout the book, the challenge is not in *how* to apply the tools, but *where*. Successful variation risk management programs include the following elements:

- **Start with the Right Set of Goals.** Variation flowdown identifies the critical customer requirements and the assembly, part, and process features that will be most likely to impact quality. By correctly identifying system KCs and creating an accurate variation flowdown, you will set the stage for working on the right problems.
- **Do Your Homework.** It is not efficient or effective to jump ahead to the solution. Teams should work through the steps of identification, assessment, and mitigation to better ensure that they pick the right problems and the right solutions.
- **Gain Consensus.** A major benefit of the identification, assessment, and mitigation procedures is that they force teams to agree on the prioritization of issues. If priorities are identified using both a quantitative and consensus-based approach, it is more difficult for teams to get sidetracked.
- **Gain Management Support.** Management support is critical for the implementation of any program. If the resources are not made available, and employees given few incentives, programs are more likely to fail.
- **Stick to Your Plan.** Once the organization has spent the time to prioritize issues, do not change it without reapplying the I-A-M procedures. It is tempting to add projects, change priorities, and veer from the original plan. Keeping the organization on a consistent path requires the dedication of both the staff and management.
- **Where Possible, Design It Right the First Time.** Don't depend on manufacturing and quality control to measure or inspect out the problems designed in. If possible, change the concept or details to make the design more robust to the existing variation.
- **Not Every Problem Is a Nail.** It is tempting to always bring out a favorite tool. But if you have only a hammer, all problems begin to look like nails. Use the I-A-M procedures to select the best mitigation strategy.
- **Look at System-Wide Costs.** Variation risk management provides the holistic view of variation that allows for trade-offs between design and manufacturing costs, between short- and long-term expenditures, and between recurring and nonrecurring costs. This holistic view will allow organizations to make the choices that are best for the organization, not the individual functional group.
- **The Goal Is the Better Design, Not the Paperwork.** VRM is a tool that should help focus the organization, not a tool for the organization to focus on.

The process of variation risk management can be aided by standard "cookbooks," consultants, and software. While all three can help to improve the execution of the VRM methodology, all should be used with caution. Hiring consultants and buying software can help you implement and manage VRM but an organization cannot expect an outside source to come in and fix its problems without involving significant in-house effort. Unless the tools and methods can be integrated seamlessly into day-to-day practices and unless incentives are put in place to ensure they are used, they will not have a long life. In short, software, consultants, and training can be useful; however, an organization must take ownership of the VRM process.

To reiterate the point made throughout this book, the hard work is not in the application of variation reduction tools, but in understanding where to apply them. The temptation will be to apply variation reduction and improvement tools where it is easy—where data is readily available and/or where the solution is obvious. Unfortunately, the projects that most need to get done are typically the most difficult to solve. Prioritizing projects is not easy; it requires a holistic view of the product and production processes to understand the true total impact of variation. The team needs to invest time and resources to properly identify and prioritize the key characteristics and select the best mitigation strategy. However, with careful planning and the VRM methodology, your organization can more efficiently and effectively apply your quality and product development resources. The result will be an increased operational efficiency allowing your organization to compete more effectively in the global marketplace.

APPENDIX A

MATURITY MODELS

Table A-1. Variation risk management maturity model for product development

Practice	Maturity Level		
	1	2	3
VRM definitions and methods (Chaps. 3–8)	The teams develop their own VRM definitions and methods. No consensus or commonality between functional groups.	Common definitions and methods are documented, but teams deviate from standards.	All teams use the same definitions and methods.
Training (Chap. 11)	No training program.	Training decoupled from process and examples are not applicable. No follow-up coaching to ensure understanding or proper application.	Programs are developed to increase skills in variation management. Training occurs just in time. Training addresses, through real examples, the problems that teams are facing. Follow-up training and coaching available to ensure proper application.
Integration into design process (Chap. 9)	VRM done at end of design.	VRM done as a separate process from design and not included in gate reviews.	IPTs use VRM throughout the design process. VRM is seamlessly integrated and is used as a criteria in the gate review.
System KCs (Chap. 3)	No system KCs identified.	System KCs identified incorrectly (for example, weight and other critical but not variation-sensitive CSRs are designated system KCs).	System KCs are identified correctly and include performance, interface, and manufacturability requirements
Variation Flowdown (Chap. 3)	Features at a piece part are identified as critical without linking them to a system KC	Part, process, and system KCs are identified but not linked through a clear hierarchy.	Coherent flowdown from system KCs through subassembly, to feature levels with clear traceability.
Ranking (Chaps. 4–7)	No ranking. Priorities are identified based on individual opinion.	Qualitative ranking system based on engineering judgment.	Ranking based on quantitative measures including historical process data, variation models, and cost data.

Table A-1. Variation risk management maturity model for product development (*Continued*)

Practice	Maturity Level		
	1	2	3
Tolerancing (Chap. 5)	Tolerances based on old designs but not checked against current capability data.	Reviews used to ensure producible tolerances, but part tolerances are not consistent with system tolerances or engineering intent.	Part tolerances are consistent with system tolerances and are achievable with current capability.
New technology (Chaps. 4–6)	New technology evaluated once in production	Potential failure modes are identified before transition to production and monitored or tested out.	Technology robustness issues are identified earlier in product development. Models are utilized to ensure robustness of the final product. Strategies to verify and control the KCs are in place.
Process capability (Appendix B)	No feedback into design.	Measurement data captured and recorded, but data is hard to find and is not used throughout the organization.	Measurement data updated regularly, easily accessed, and used by design to verify capability.
Uncertainty reduction (Chaps. 4–6)	Uncertainty never reduced. Problems identified in production.	Modeling is used at the end of the design process.	Early and continued use of prototypes and variation models to reduce uncertainty.
Mitigation selection (Chap. 8)	No systematic evaluation and selection of mitigation strategies.	Mitigation strategies are applied late and are heavily weighted toward QC.	Mitigation methods are chosen based on cost and benefits. The team attempts to fix problems as early as possible.
Robust design (Chap. 8)	Robust design not used.	Reactive changes are made to areas that are found not producible during transition to production.	The design is made robust. High-risk areas are identified and mitigated as early as possible.

Quality control planning (Chap. 7)	Quality control plans developed separately from the VRM process.	Quality control is used as the default mitigation plan.	Special quality control only used where design could not be made robust enough. Quality control plan is designed such that defects are detectable and diagnosable, and the resources are used in an efficient and effective manner.
Suppliers (Chap. 10)	Drawings and designs handed over the wall.	Suppliers are brought in at the end of the design to verify the KC capability.	Suppliers are integrated with the IPT's to evaluate KCs. They make suggestions where the design may not be robust and where changes will have a significant impact on cost quality and performance.
IPTs (Chap. 10)	Individual functional groups responsible for VRM.	Cross-functional teams formed when a problem arises.	IPTs are formed proactively to work on VRM and work in a fire prevention mode.
Management support (Chap. 11)	No management support for VRM process.	Management supports teams to use VRM, but VRM is ignored if larger problems arise. Management rewards putting out fires.	Management understands the need for VRM and advocates and helps facilitate its application. Management rewards good application of VRM and preventing fires. Management is actively involved in gate reviews.
Documentation (Chap. 11)	No documentation of flowdowns, assessment, or mitigation strategies.	Information is scattered across a variety of functions with no centralized source.	Identified KCs are well documented and traceability is established. All KCs and supporting documentation are updated regularly and available for future designs.

Table A-2. Variation risk management maturity model for production

Practice	Maturity Level 1	Maturity Level 2	Maturity Level 3
VRM definitions and methods (Chaps. 3–8)	The teams develop their own VRM definitions and methods. No consensus or commonality between functional groups.	Common definitions and methods are documented, but teams deviate from standards.	All teams use the same definitions and methods.
VRM process initiation (Chaps. 3–8)	No VRM program. Problems are fixed as they arise.	Problems and their root causes are identified as they appear.	The KCs are identified. Projects are prioritized to continually reduce cost.
Variation flowdown (Chap. 3)	Features at a piece part are identified as critical without linking them to a system KC	Part, process, and system KCs are identified but not linked through a clear hierarchy.	Coherent flowdown from system KCs through subassembly, to feature levels with clear traceability.
Cost of variation (Chap. 6)	No cost calculated.	Ranking based on cost to single functional areas. Data sources from all functional groups are not integrated.	KCs are ranked according to the total cost of variation including customer costs, warranty, scrap, rework, repair, inventory, and capacity.
Mitigation selection (Chap. 8)	No coherent process.	Mitigation strategy selected based on qualitative assessments.	Mitigation strategy selected based on cost and benefits. The team attempts to fix problems as early as possible.
Quality control maturity (Chap. 7)	Detectable.	Diagnosable.	Efficient.
Suppliers (Chap. 10)	Relationship based on inspection of parts.	Suppliers brought in only if a problem occurs.	Suppliers are integrated into proactive VRM.
Management support (Chap. 11)	Management encourages the use of VRM, but resources are not properly allocated.	Management encourages the use of VRM, but focus on process is abandoned when a fire arises.	Management understands the need for VRM and advocates and helps facilitate its use within the organization.
IPTs (Chap. 10)	No teams used.	Formed when there is a problem in production. Formed for fire fighting.	Formed proactively to work on cost reductions. Work in a fire prevention mode.

APPENDIX B

PROCESS CAPABILITY DATABASES

This appendix provides details on the issues and challenges of implementing a process capability database. Using process capability is critical to the VRM methodology and many organizations are establishing a process capability database. However, implementation can be fraught with unexpected issues. This chapter provides tools, techniques, technology and suggestions to help team efficiently implement a process capability database.

Process capability is the measure of how much variation a manufacturing process generates in relation to the target tolerance (Chapter 2). Understanding process capability is critical to the VRM methodology in that it is used to validate manufacturing processes before production. More important, it supports a data-driven approach to assessment. During production, manufacturing process capability databases can be used to compare capability across an organization or to help in selecting mitigation strategies. Process capability is used in two ways:

1. **Tolerance Analyses.** Process capability data can be used to populate variation models to predict the variation in system KCs.
2. **Tolerance Allocation.** If tolerances are correctly allocated to the individual part KCs, the process capability database can be used to validate whether or not individual tolerances are achievable.

Typically, one of two methods is used to obtain process capability data. The first method depends on direct communication between design and manufacturing. In this method, engineering typically literally asks manufacturing, "Can you achieve this tolerance on this design?" Manufacturing then uses its own internal data resources and expertise to answer the question. The second method depends on a database that contains information on manufacturing process capability and is accessible to the design and production communities. The database is queried using part, feature and process characteristics and the data that most closely matches the new design is returned. Each method has its benefits and shortcomings. The first is time consuming but typically results in the right data being presented. The second is theoretically less time consuming but has many technical issues surrounding the implementation and use of the database.

The second method, process capability databases, is becoming the approach of choice for many companies. The rest of this chapter will review the challenges and tasks in implementing and using process capability databases. The next section provides background on process capability databases and subsequent sections include suggestions for implementation strategies, technologies, and processes. Unfortunately, no viable commercial software has been developed that can be used as a starting point for any implementation.

B.1. BACKGROUND ON PROCESS CAPABILITY DATA

While using process capability is critical to variation risk management, the design and development of a process capability database is not a simple task. Implementation requires understanding how the database will be structured, what data will be included, how the database will be used in the design process, and how it will be supported. This section deals with the importance of using process capability data and introduces the challenges facing any implementation.

B.1.1. Importance of Using Process Capability Data

Traditionally, tolerance specification has been unidirectional—that is, the designer specifies tolerances and manufacturing is expected to meet them. However, in the new paradigm of design for manufacturing (DFM) and concurrent engineering, communication with manufacturing is necessary to ensure that specified tolerances are consistent with manufacturing's capability. By ensuring producibility before a design is released to manufacturing, expensive redesign, rework, and customer dissatisfaction can be avoided.

Figure B-1 shows two approaches to product design and manufacturing process selection. In the process at the bottom of the figure tolerances are selected based on history or on engineering judgment. In either case, tolerances are set without understanding how variation in the part KCs rolls up and impacts variation in system KCs. After the design is completed, manufacturing selects its processes based on what is needed to achieve tolerances. Because

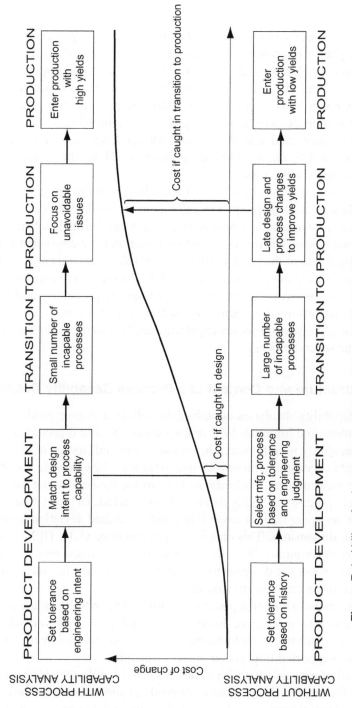

Figure B-1. Utility of using process capability data to validate designs prior to transition to production.

of mistrust between design and manufacturing, design engineering often artificially tightens tolerances to ensure that they get manufacturing's attention. Knowing tolerances are not "real," manufacturing does not select processes to meet the actual tolerances, but rather the assumed real tolerances. When a process is found not to be capable in production, one of three expensive responses occurs. First, significant engineering work goes into opening tolerances, and drawings must be changed. Second, the manufacturing process must be changed to reduce variation. Third, the design must be altered to make it more robust.

In the second process at the top of Figure B-1, manufacturing process capability data is used in product development to validate processes *before* transition to production, and necessary changes can be made when the cost is significantly less. Tolerances are set such that they are consistent with the engineering intent of the product (the part KC tolerances are consistent with the system KC tolerances) and are producible. As the design process continues, the design is created to match both customer requirements and process capability. In transition to production, the team can focus on the unavoidable or unpredictable issues. Process capability databases enable organization to practice the second process.

B.1.2. Structure and Content of a Process Capability Database

Process capability databases contain data collected during production. As parts are measured for either SPC or inspection, the data is recorded and the process capability is computed. The measurement and storage of process capability data is becoming more widespread through the extensive use of SPC, automatic measurement machines, and electronic data collection methods.

A typical process capability database record includes a target mean, an actual mean, an upper limit, a lower limit, and a standard deviation for critical component dimensions. This data is used to calculate yield, DPM, and C_{pk}. In addition, descriptors (also called an index) of the part and process used are included to enable the designer to search the database for the surrogate data that most closely matches the new design.

It is important to note that process capability data in the database provides only a *surrogate validation*. Every design and part is slightly different. Changes in geometry, manufacturing process, and materials can have dramatic and unexpected impacts on process capability.

The goal of process capability databases is to identify the closest possible surrogate for the new design. The team needs to understand how the new design varies from the previous designs and the potential impact on process capability. As with mutual fund financial disclosures, previous performance does not guarantee future performance.

B.1.3. Difficulties in Implementing Process Capability Databases

In general, when organizations were questioned about implementing process capability databases, their first reaction was, "This should be easy." They believed that if data was simply put online in a structured format, engineering would be able to access and use the data to ensure producible designs. However, this idealistic view is far from the truth of what is involved in creating a usable process capability database (Tata and Thornton, 1999). There are many issues involving indexing methods, user interfaces, usability, using surrogate data, and keeping data updated and correct.

Many companies have made unsuccessful attempts at implementing process capability databases that can be used in design. Data is collected and used to monitor manufacturing but is not used during product development. Figure B-2 shows a cause and effect diagram developed to explain the failure of companies to use process capability databases in design (Tata and Thornton, 1999). The ability to use process capability data in product development is subject to four factors:

1. **The Right Structure.** The organization must have a common indexing scheme for storage to allow for multiple sites to both contribute to and access the data. A common scheme is needed if the data resides in multiple databases. The user interface must be interactive and easy to use.
2. **The Right Data.** The organization needs to populate the database with up-to-date data that accurately represents the current process capability. In addition, the data must reflect the data needed for new product design. Ideally, the supplier data should also be included.
3. **The Right Management Support.** Management must provide incentives for using process capability data and resources for its implementation and maintenance.
4. **The Right Usage.** The IPTs must effectively use the data to benefit new product development. Ideally, process capability data (PCD) should be integrated with other information and design systems to allow for seamless evaluation of new product designs.

The next four sections outline issues involving the four factors.

B.2. THE RIGHT STRUCTURE

The structure of a process capability database will have a significant impact on the quality of the data, the usability of the system, and the long-term

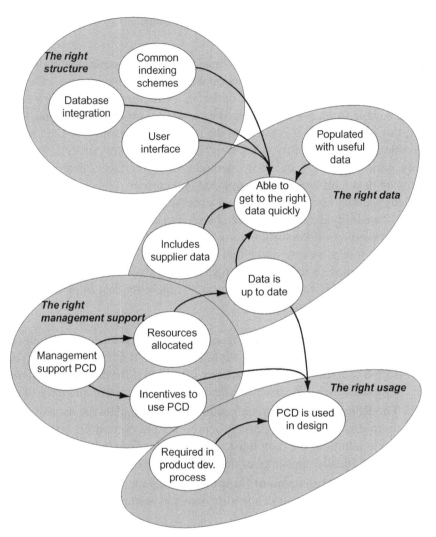

Figure B-2. Key drivers for process capability data usage (Tata and Thornton, 1999). Reprinted with the permission of ASME.

benefits accrued. Designing the structure of a process capability database involves understanding the following factors:

- The methods of indexing the data
- The database approach
- The user interface

The following sections review each of these points.

B.2.1. Designing the Indexing Scheme

Once the data is collected by manufacturing and/or quality control, the designer needs a way to access it. In the past, data was often indexed according to drawing number. To look up process capability for a given feature, one would need to know which drawing had a similar feature; this made searching for surrogate data difficult. To alleviate this problem, several companies started to index data based on descriptors rather than part numbers.

When a team asks, "What is the standard deviation and mean shift for a drilled hole?," the answer they often receive is, "It depends." The search for the correct capability is not trivial, because many factors can influence the final process variation, and these factors need to be recorded alongside the raw data. A number of different descriptors can be included (Hanson, 2001; Kern, 2003; Thornton, 2001. Reprinted with permission of Kluwer Press) (the example of a hole is used to illustrate each descriptor):

- **Process:** Holes can be created using a variety of processes. Each process will introduce different amounts of variation.
- **Material:** Given the same process, some materials will introduce more variation than others.
- **Feature and Dimension:** The process variation for hole drilling will depend on the aspect of the feature being measured. For example, the hole depth may have less variation than the diameter. In addition, variation in pocket depth will be different from variation in hole depth. The feature type and the dimension being measured must be specified.
- **Feature Size:** The process variation in holes with larger diameters may be larger than that for holes with smaller diameters.
- **Machine:** Many designers will set tolerances assuming the parts will be made on the highest capability machine. However, the most capable machine may have limited capacity. Consequently, overflow parts might be made on older, less capable machines. Ideally the least capable machine that still delivers the acceptable quality should be chosen.
- **Manufacturing Site:** The same design can be made at multiple locations with the same equipment with varying success. Different sites can have better control, maintenance, and operating procedures that may result in different process capabilities.
- **Tolerance:** The tolerance specified on the part drawing may change the manufacturing process and the settings used. For example, for tight-tolerance surfaces, a NC path planner may specify a slower speed cut speed. However, tighter tolerances typically come at a cost of increased labor content, manufacturing processing cost, and/or cycle time.

- **Operator:** Machine operators can influence process variability based on skills and experience.
- **Part Characteristics:** A feature's location in a part can influence process variation. For example, more variation will be introduced when drilling a hole in a thin part than in a thick part. Fixtures and postprocessing (such as polishing) will also influence the final variation.
- **Part Volume.** In low-volume production, highly-trained operators in a job shop environment may make parts, whereas in a high-volume shop, automated equipment may be used. Using automation in high-volume production may standardize processes, but the job shop might be more capable of controlling variation on difficult-to-produce parts.
- **Stock.** The same material may come in a variety of stock types. For example, aluminum could be delivered in billet form or forged in a near net shape. The machining on both will cause different variations because of the differences in material properties.

Several descriptors are numerical (e.g., tolerance, number of parts), but others are based on a set of possible values (i.e., material, manufacturing process, feature, machine). Still others are too general to be recorded in a structured format (e.g., feature/part characteristics). For any indexing scheme there is always a trade-off between the specificity of a descriptor set and the overhead involved in managing the data. It will never be possible to completely capture all the factors that influence the capability in a structured data set; to address the more complex subtleties the manufacturing experts may need to provide input on the process capability.

An indexing scheme is a systematic method for describing typical descriptor combinations. The standard descriptors—material, feature, stock, and processes—are given unique codes, so that a combination of feature, material, manufacturing process, and dimension codes uniquely identifies each data record. Additional variable data such as feature size, tolerance, and batch size are also included. Typically, each indexing code has a hierarchical structure used to facilitate coding and retrieval (Kern, 2003; Tata and Thornton, 1999).

Figure B-3 shows a sample indexing scheme for plastics and metals.

- The **materials** are grouped into classes (e.g., thermoplastics and metals). Each of these has subclasses (e.g., steel and iron). Each subclass has specific types and so forth. The indexing scheme could include details down to the specific alloy mix and processing for the various subclasses of materials.
- The **stock** types can include a number of different shapes and sizes.

MATERIAL

- 1.0 Thermoplastic
  - 1.1 ABS Polymer
  - 1.2 Acetal
  - 1.3 Acrylic
  - 1.4 Elastomer, TPE
  - 1.5 Ethylene Vinyl Acetate
  - 1.6 Nylon
  - 1.7 Polycarbonate
  - 1.8 Polyester, TP
  - 1.9 Polyethylene
- 2.0 Metals
  - 2.1 Steel
    - 2.1.1 Stainless
    - 2.1.2 Low Alloy
    - 2.1.3 Carbon
    - 2.1.4 Tool
  - 2.2 Iron
    - 2.2.1 Alloy Cast
    - 2.2.2 Ductile
    - 2.2.3 Gray Cast
    - 2.2.4 Malleable
    - 2.2.5 White Cast
  - 2.3 Aluminum
  - 2.4 Copper Alloy
  - 2.5 Titanium

PROCESS

- 1.0 Thermoplastic Molding
  - 1.1 Injection
  - 1.2 Extrusion
  - 1.3 Thermoform
  - 1.4 Roller
  - 1.5 Transfer
  - 1.6 Compression
  - 1.7 Vacuum
  - 1.8 Dipped
  - 1.9 Pressure
  - 1.10 Rubber
- 2.0 Machining
  - 2.1 Boring
  - 2.2 Milling
    - 2.2.1 End of Cutter
    - 2.2.2 Side of Cutter
    - 2.2.3 Kellering
  - 2.3 Sawing
  - 2.4 Shearing
  - 2.5 Turning
- 3.0 Forming
  - 3.1 Stretch
  - 3.2 Brake
  - 3.3 Rolling
  - 3.4 Swagging

STOCK

- 1.0 Angle
- 2.0 Bar
  - 2.1 Flat
  - 2.2 Hex
  - 2.3 Round
  - 2.4 Square
- 3.0 Casting
- 4.0 Coil
- 5.0 Extrusion
- 6.0 Pellets
- 7.0 Plate
- 8.0 Strip
- 9.0 Tube
  - 9.1 Round
  - 9.2 Square
  - 9.3 Rectangle

FEATURE

- 1.0 Chamfer
  - 1.1 Angle
  - 1.2 Depth
- 2.0 Hole
  - 3.1 Depth
  - 3.2 Diameter
- 3.0 Open Profile
  - 3.1 Contour
  - 3.2 Position
- 4.0 Pocket
  - 4.1 Length
  - 4.2 Depth
  - 4.3 Width
  - 4.4 Position
- 5.0 Rib
  - 5.1 Height
  - 5.2 Depth
  - 5.3 Position
  - 5.4 Thickness

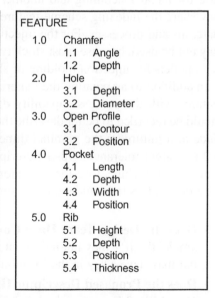

Figure B-3. Sample indexing structure (Thornton and Tata, 2002). Reprinted with permission of Cambridge University Press.

- The **processes** includes molding, machining, and forming. Typically metals are not manufactured using a molding process or plastics by machining, but there may be some crossover.
- The **features** are the individual features and dimensions where the measurement occurs. The list of features and dimension can be generated by reviewing the GD&T standards.

For the indexing scheme in Figure B-3, if a machinist measures the depth of a hole by milling with the end of the cutter in an aluminum extrusion 6AL6V:, the index would be M2.3 S5.0 P2.2.1 F3.1.

Indexing is not without problems, however. At one company they had generated an indexing scheme for all of its manufacturing processes. It could encode 52 million possible combinations, but only 50,000 of the combinations (0.1 percent) were feasible, and many feasible index combinations were not populated. An indexing scheme must be developed with an appropriate specificity of coding. Codes that are too detailed will result in a sparse data set; codes that are too general may be difficult to interpret.

One way to avoid the sparse data set problem is to not implement a single monolithic database but to create a series of separate databases for each family of processes. For example, the organization may decide to have one database for metal machining and another for plastics forming. The databases may share the indexing scheme features but will have different indexing for materials and processes. For the injection molding, the stock material index may not be used, because most stock types will be determined by processing (e.g., pellets for injection molding or sheets for thermoforming).

In addition to associating the material, stock, manufacturing process, and features with the process capability data, other variable and attribute data should be recorded to better describe the data. This additional data should include at a minimum the nominal dimension of the feature being measured and the target tolerance. Other descriptors can include those from the preceding list (machine, location, operator, batch size, etc.). When determining what descriptors to include, the implementation team should ask:

- **Does the Design Team Have Control over This Descriptor?** For example, the IPT may not have control over which operator creates a part, but may have a say in the location at which the part is built.
- **Does the Proposed Descriptor Have a Major Impact on the Process Capability?** For example, batch size may have a large impact on long-term capability for the same process at the same location if the setup introduces a large amount of variability.

• **Can the Descriptor Be Easily Formalized?** If there are clear attribute lists (i.e., machine number) or the parameter is variable (i.e., tolerance or dimension), then the descriptor can be easily included. If the descriptor is less concrete (i.e., type of part), then it may not be reasonable to include the descriptor because it is so difficult to formalize.

B.2.2. Choosing the Database Implementation Approach

Once the indexing scheme and data structure have been chosen, the database implementation approach should be determined. Most companies today do not use a single manufacturing site to produce their parts. There are redundant manufacturing facilities within the organization, and the organization can make use of supplier production capacity as well. The implementation team should determine how these various database sources can be integrated to enable the design team to understand the corporate capability, not just the site capability.

A number of approaches exist for implementing a process capability database that spans multiple sites. A number of questions must be asked:

• **Is It Necessary to Have Real-Time Data?** Does the database need to reflect parts being built today, or can older data be used?
• **How Error Prone Is the Data?** If the data is error prone, using a real-time database may be difficult, and the data may need to be cleaned up before being posted to a central location.
• **How Common Are the Data Formats and Software across the Organization?** If the data formats and software are diverse and incompatible, integrating locations in real-time may be difficult; most of the implementation time will be spent getting the diverse databases to be compatible.

There are two approaches to integrating the data across the organization (Thornton, 2001b. Reprinted with permission of Kluwer Press):

1. **Real Time:** In this approach, multiple data sources from multiple sites can be accessed in real time, and users will have access to the most up-to-date data. However, implementing a real-time database requires a significant amount of time and information technology skills. Each site will need to add an indexing scheme to its database to ensure conformity of data, and automatic data checkers must be employed to remove any errors in the data.
2. **Meta-Database:** In the meta-database approach, data from each site is uploaded regularly to a centralized location. During the uploading

process, the data is cleaned up, indexed, and put into a common format. This can be done automatically or manually. Data in the meta-database will lag behind the real-time data by up to six months. If this lag is not a problem, implementation should be significantly easier than for real-time databases.

B.2.3. Creating the User Interfaces and Data Analysis

Ideally, process capability databases should be integrated with both the CAD systems and any variation analysis tools. However, this is often not possible for two reasons. First, the CAD systems and variation analysis tools are not capable of intelligently identifying all descriptors necessary to find the correct surrogate data. For example, the CAD model does not "know" what a boss or a rib is. Second, there is a large technical hurdle to integrating the two programs. In the short term, most process capability databases will need to be searched by the user and manually transferred to any analysis package.

The user interface of the process capability database is critical to the usefulness of the tool. If the user interface is difficult to use, it will reduce the chance that the design IPT will use it. There are two parts to the user interface: first, the system by which the operator inputs the data and descriptors, and second, the ways the data can be interpreted once it is accessed for analysis. To obtain surrogate data a query is sent to the database for a specific process, feature, and material combination; the database will return no, one, or multiple sets of data.

The simplest user interface (Fig. B-4) uses pull-down menus to specify the material, stock, manufacturing process, and feature. In addition, the target dimension and tolerance width are also entered.

The pull-down menu user interface can be made more intelligent in four ways:

1. **Restricting the Pull-down Menus to Valid Combinations.** For example, when a metal material is selected, only the machining processes would appear in the pull-down menu.
2. **Restricting the Pull-down Menus to Those Combinations That Are Populated with Data.** By sequentially searching the database with each selection, it is possible to dynamically update the other pull-down menus to show only the populated indices.
3. **Allowing Users to Select Ranges of Dimensions and/or Tolerance Widths.** For example, there may be data for diameters of 3.1 and 3.3 but not 3.2. By picking ranges, users can obtain surrogate data that is relatively close to the target value they are searching for.

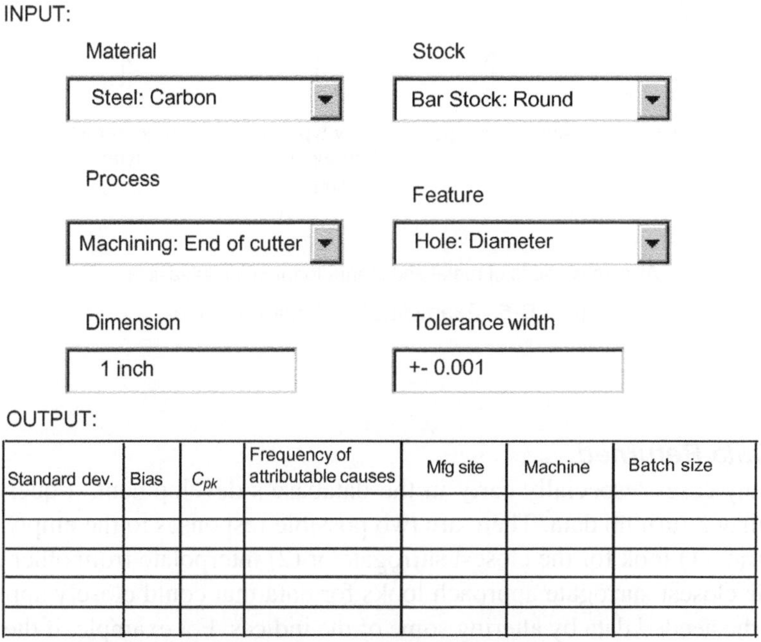

INPUT:

Material

| Steel: Carbon ▼ |

Stock

| Bar Stock: Round ▼ |

Process

| Machining: End of cutter ▼ |

Feature

| Hole: Diameter ▼ |

Dimension

| 1 inch |

Tolerance width

| +- 0.001 |

OUTPUT:

Standard dev.	Bias	C_{pk}	Frequency of attributable causes	Mfg site	Machine	Batch size

Figure B-4. Sample user interface.

4. **If an Empty Combination Is Selected, Enabling the System to Intelligently Search for a Set of Data Points That Are Close.** For example, if one alloy of aluminum does not have data for milling holes, another might.

When searching the database, the user must understand the difference between an "empty" index and an infeasible one. An empty index is one where the organization has not collected data on that specific material, stock, manufacturing process, and feature combination. An infeasible index is one where there should not be any data. For example, you are not going to injection-mold a titanium extrusion.

To differentiate between infeasible and empty indices, the implementation team can list rules of infeasible combinations. These rules can be used when searching the database. An asterisk represents a wild card. For example, the rule that states that no metal can be injection molded would be described as shown in Fig. B-5.

There are two possible outcomes of a search on a process capability database. In the first, no data is returned because no similar combination has been produced before. In the second, multiple records are returned. In this case, all records may indicate a capable or incapable process or may return inconclusive data.

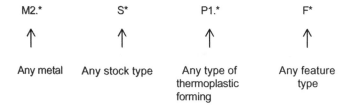

All combinations of metal and thermoforming are infeasible

Figure B-5. Example of an infeasible index.

No Data Returned

In many cases, especially early in the database's development, a query for data may return no data. There are two possible responses to the empty data scenario: (1) look for the closest surrogate or (2) interpolate from other data.

The closest surrogate approach looks for data that could closely approximate the needed data by altering some of the indices. For example, if the user requests the capability for a hole of diameter 3 inches, created by an end-of-cutter in an aluminum bar, the system could return a number of similar data points including the same hole in an aluminum billet or the same hole in an aluminum bar of a different alloy.

In another case, similar data for the same index may be available but for different dimensions and/or tolerances. The available data can be used to interpolate and estimate the unpopulated index. For example, a designer is interested in understanding the capability of producing a diameter of 3.25 and is returned the data shown in Table B-1. By plotting the data points and doing a linear regression (Fig. B-6), it is possible to estimate the capability of producing a diameter of 3.25. The result of the linear regression is:

$$C_p = 0.105D + 1.0, \qquad R^2 = 0.906 \qquad \text{(B-1)}$$

Table B-1. Sample data

Diameter	C_p
1	1.15
1.5	1.1
2	1.23
2.5	1.245
2.75	1.295
3.5	1.34
4	1.45

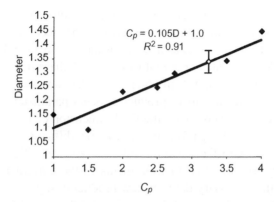

Figure B-6. Using linear regression to estimate capability.

The model (Eq. B-1) predicts a C_p of 1.34 with an R^2 of 0.906 (the model explains 90 percent of the observed variation in C_p). The standard deviation around the line is 0.039, which means there is a 69 percent chance that the capability will fall between 1.30 and 1.38.

Too Much Data Returned

A problem also occurs when multiple and conflicting process capabilities are returned. If all capabilities are either acceptable and/or all capabilities are un-acceptable, it is clear that either the design is definitely okay or definitely not okay. If some of the records indicate that the manufacturing process is capable and some that it is not, then the team needs to work to understand why the same process has different capabilities. Some possible causes for differences in capability include:

- **Differences in Machines, Operators, or Locations.** Two different machines or the same machine run by different operators can have different capabilities.
- **Differences in Part Geometry.** For example, drilling holes in thin parts generates more variation than drilling in thicker parts.

By analyzing the other descriptors on the data sets, it may be possible for the user to deduce the cause of the differences. In most cases, a meeting with the manufacturing organization may be the most expedient way of determining which of the runs is representative of the capability for the new design.

B.3. THE RIGHT DATA

Another necessary but not sufficient condition for a successful database is that it must be populated with useful data. This obvious requirement is often overlooked.

In a typical scenario, the factory floor first manufactures a part or assembly and measures the key variables. The measurements are recorded along with a set of descriptors. The data starts as a series of measurements taken on every part or on a sample of products. For real-time control, these measurements are used in statistical process control; in other cases, they are used as part of the inspection or buy-off of the part or product. Process parameters may also be monitored by production rather than the final dimension or feature. This data may not be useful for design because it can be difficult to translate the processes parameters into the capability of achieving specific tolerances. It may be necessary to add some measurements to the current measurement plans to provide the necessary data for new product design.

To present data in a form useful to designers, the raw data is typically aggregated into production runs that contain the mean, standard deviation, and number of samples. The runs can be analyzed at a number of different levels.

- **Short-Term Capability.** This is the capability of the process assuming the manufacturing process is stable. The capability includes both the standard deviation and the mean shifts and is typically measured over a short period of time.
- **Long-Term Capability.** This is the capability measured over a longer period of time. The long-term capability can have greater variation than the short-term capability: capability can change between shifts, is usually poorer during setup, and degrades due to machine wear.
- **Attributable Causes of Variation.** The process capability can degrade suddenly due to assignable causes of variation. The frequency of these causes, the duration before they are removed, and the magnitude of degradation should be analyzed.

In addition, the data set may include a measure of gage repeatability and reliability. The gage R&R measures variation that is introduced by the measurement system. In some cases, the gage variation can represent a significant amount of measured variation. When presenting the data to the user, it is necessary to identify whether the data includes or excludes the gage variation, whether the data is short term or long term, and whether the assignable causes have been removed from the data sets.

The raw data can have a number of flaws. Errors in measurement and data entry can corrupt the data. In addition, selective entering of data can skew the true process capability. Selective data entering can happen when a person chooses to enter only the data that shows stable performance and throws away the data taken when the manufacturing process was out of control. For each data source, the data integrity must be checked. This analysis must occur wherever data is generated.

In most manufacturing organizations, many sites are involved in production. The designer may not know which site will build the part. Therefore, a survey of all manufacturing sites should be conducted, and the data collection methods, existing data, and databases should be catalogued for each site. When selecting which data to include in the centralized database the implementation team should take into account the organization's manufacturing strategy. For example, if one of the manufacturing plants has antiquated equipment that will not be used for future production, its data should not be included in the database.

In situations where the organization designs parts that are outsourced, a supplier's capability must be understood. In general, suppliers are wary of sharing this data, although key suppliers with whom the organization has had a long relationship may be more willing. The meta-database approach is more compatible with including supplier data than is the real-time approach.

There are several sources for supplier data:

- **Product Data Sheets.** These often provide a summary of the process capability of key features and specifications.
- **Quality Control Reports.** Some companies require their suppliers to provide statistical data about parts they are delivering.
- **Incoming Inspection.** If parts are inspected when they arrive, this data can be collected.
- **Data Direct from the Supplier.** Some suppliers may be willing to share their internal process capability data. However, this is rarely the case. There is a fear among many suppliers that the data will be used to compare suppliers and/or renegotiate contracts.

B.4. THE RIGHT MANAGEMENT SUPPORT

A group may design the most useable database but without management support its success may be limited. Management must support the development and deployment of a process capability database in two key ways:

1. **Resources.** The development of a process capability database will require both one-time resources to design and build the database and continual resources to support its upkeep.
2. **Incentives.** Product development team leaders need to hold their IPT members accountable for using the data during product development.

During review processes, management must ask questions about how a capability was verified and what data was used.

B.5. THE RIGHT USAGE

Once data is indexed, collected, cleaned and entered into a database it can be used in several ways to:

- **Provide Data for the Assessment Phase.** The standard deviations and mean shifts can be used to populate variation models.
- **Validate Individual Tolerances.** The process capability database can be used to assess the capability of achieving individual tolerances.
- **Determine the Best Manufacturing Process.** When choosing a process, location, or machine to use for production, the database can be used to help select the equipment that most closely matches the needs of the product.
- **Determine the Most Cost-Effective Tolerance.** By adding standard costs and capacity to the database, the design team may be able to trade off the cost of the manufacturing process against the tolerance it can deliver. This will allow designers to determine how to allocate tolerances to minimize total product cost.
- **Ascertain the Tightest Possible Tolerance.** The data provides a list of tolerances that have been achieved for various manufacturing processes. If the tolerance set by design is tighter than those included in the database, manufacturing engineering must be consulted to understand the ability of the process to achieve the tighter tolerance.

For all the benefits a process capability database affords it does not or should not replace the manufacturing engineer. As has been pointed out, data returned by the database will not always answer the question, "Is the process capable?" There will be a small subset of tolerances that require detailed reviews with manufacturing.

The team should focus on the difficult 10 percent of cases and use the database to support the validation of the easy 90 percent. Figure B-7 shows the flow of how to use the database during product development.

The part KCs associated with the highest-risk system KCs should be the first tolerances checked using the database. If multiple surrogate data sets are returned, the sets should be analyzed to determine if the data sets all indicate a capable process, if they all indicate an incapable process, or if the situation is unclear. If the answer is not clear, then the design team should work with the manufacturing engineering staff to understand what process capability is

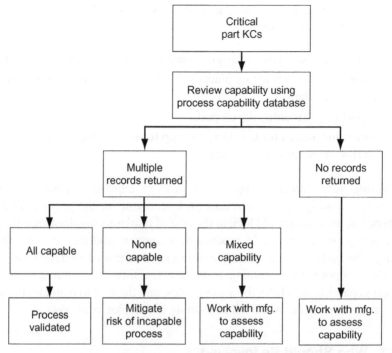

Figure B-7. How to analyze process capability.

likely to be. If no data sets are available, the team should work with manufacturing to determine what the likely process capability will be.

As was pointed out in the introduction to this chapter, the utility of a process capability database is to identify, in a systematic way, previous designs and processes that can provide a surrogate for the new design. By closely matching the descriptors, the team can identify possible matches. However, the database users must not assume that the data returned is a guarantee of the capability of the same process on the new design. Small differences in geometry or the manufacturing process used can result in different capability.

B.6. IMPLEMENTATION OF A PROCESS CAPABILITY DATABASE

The previous two sections gave detailed descriptions of the technical aspects of implementing a process capability database. This section will outline the major steps and what functional groups should be involved in designing and implementing a process capability database.

A single group should take ownership of the database. This group can either be from a single functional group or be an IPT. The first step in developing the database is to gain management support. Management must be willing to do the following:

- Provide the resources to design, implement, and maintain the database
- Require the manufacturing plants to be cooperative and to supply timely and accurate data
- Support any standards developed
- Provide incentives for the design group to use the data within the variation risk management program

Gaining management support for process capability databases often hinges on showing a positive return on investment for the project. To argue the benefits, the team should look at the cost of retolerancing drawings, the cost of changing processes late in product development, and the cost of reduced quality. The budget for the project should include the time to design the product, the computer programming resources, the time to collect and analyze process capability data from the various sites, and the cost to maintain the database on an ongoing basis as well as training and deployment costs.

B.6.1. Who Should Be Involved

The functional groups involved in designing the database should include:

- **Owners of the Database.** This is ideally a cross-functional group that has a stake in getting the database to work correctly.
- **Information Technology.** Often, internal information technology groups will take responsibility for implementing the database and user interface. They will be able to provide the robust and maintainable software; however, they will require a clear set of requirements from the rest of the team for both the data structure and user interface.
- **Manufacturing Engineering.** The manufacturing engineers typically own the process capability data and are responsible for assessing capability. Their way of finding data, analyzing it, and determining the most appropriate surrogate data should influence the design of the indexing schemes and the user interface.
- **Users of the Database.** Users will include the design community. They should be interviewed to understand what their needs are and what information they have available when they need to search the database. In addition, they should evaluate any prototype software.
- **Manufacturing Plants.** The manufacturing plants will provide the data to populate the database. The plants should be included to assess the current state of the data, how it is indexed, and how valid it is.

B.6.2. What Decisions Should Be Made

A number of decisions will need to be made before the database design and implementation begins. These decisions include the following:

- **What Manufacturing Areas Will Be Included.** In addition, is a single database or are multiple databases going to be developed. For example, a company with expertise in metals machining and composites may want to build two databases. It is recommended that the team start with one database, implement it, and then use it and the lessons learned as a template for future implementations.
- **The Indexing and Descriptor Schemes.** Determining these schemes can be done as a group effort. The indexing scheme is critical to identifying the correct surrogate data. The following are steps that can be taken to help generate the indexing and descriptor scheme:
  —Review drawings of typical parts and collect a set of common features, manufacturing processes, and materials.
  —Review the common dimensions used. The GD&T standards can provide a list of possible dimensions.
  —Review manufacturing processes and collect the features created and materials used.
  —Review all quality, measurement, and KC plans and documents. This will give the team an understanding of what the most critical processes are.
  —When designing the indexing scheme, use a standard 1./1.1/1.1.1 numbering scheme. This will allow for later expansion.
  —Identify the numerical or a enumerable data that can be associated with each measurement (e.g., dimension, tolerance, manufacturing process parameters, date, machine, operator, and location).
- **The Database Approach.** The team needs to select between the meta-database and the real-time database approaches.
- **The Manufacturing Data to Be Included.** It is worth doing a data inventory to determine what capability data is available. Additional data may need to be collected to populate the critical areas of the database.
- **Method of Data Distribution.** The data can either be maintained in local databases or made available to other engineers via a web interface.
- **Who Has Access.** The organization may decide to limit access to the database to only a select set of people. If the database includes supplier data, the suppliers may be sensitive to access issues.
- **How Modifications Will Be Handled.** Policy and procedures need to be created are for modification and/or expansion of the indexing scheme or the database.

- **Database Integration.** How will using the database be integrated into the product development process should be determined by the VRM and process capability implementation teams.
- **How the Usage and Success of the Database Will Be Tracked.** This can include the number of times the database is accessed, how many tolerances were checked, and, during transition to production, how many incapable processes were identified that had been previously validated.
- **Maintenance.** What long-term support is going to be provided? How will the data be maintained and by whom?

B.6.3. Implementation Steps

Once the preceding decisions have been made and the resources identified to build and maintain the database, the team should execute the following steps to develop the database:

- **Finalize the Indexing Scheme.** Once the scope of the database is determined, the indexing scheme should be developed.
- **Collect Data and Clean It.** The data to be included in the database should be collected and cleaned to remove errors and/or erroneous data.
- **Develop a User Interface.** Based on the requirements of the organization, the indexing scheme, and inputs from users, the user interface should be developed.
- **Specify and Implement Software.** The actual tools should not be implemented until a clear set of specifications, needs, and structures are determined.
- **Develop Training.** A training program will be needed to deploy the system.

B.7. SUMMARY

The process capability database is critical to the VRM methodology because it provides key inputs into the assessment phase: (1) providing quantified measures of capability and (2) identifying the critical few manufacturing processes that need to be validated using manufacturing expertise. While many companies are considering implementing or have implemented databases, the development of a process capability database is not as simple as just accessing measurement data online. Significant thought needs to be put into designing the database, populating it, and using the data.

APPENDIX C

OTHER INITIATIVES

Product development and manufacturing programs are typically awash with acronyms and buzzwords: Six Sigma, robust design, design for manufacturing, and so on. Each of the new tools and methods has the goal of encouraging good design, good behavior, reduction in waste, and increases in operational efficiency.

No one of these initiatives will provide the complete solution to a company's ills. Each tool has its own strengths and weaknesses. Often organizations pick a "solution," whether it is Lean or Six Sigma, and end up believing that it can solve all problems within a company. However, all losses and operational inefficiencies cannot be understood by single teams, nor can one solution address all needs. One should keep in mind the poem by John Godfrey Saxe entitled "The Parable of the Blind Men and the Elephant" (see text box). In the same way, each functional group or IPT can only "touch" one part of an organization's operational inefficiency and given freedom will pick the best tool to address that aspect. The challenge arises when someone wants to apply a tool optimal for the trunk to a leg. Figure C-1 shows the "elephant" of operational inefficiency.

When implementing any initiatives, several things must be kept in mind.

- **Buying a Solution Does Not Mean You Will Get Results.** Hiring consultants and doing extensive training will not guarantee you results. The organization needs to make changes and follow up on improvements to see the long-term benefits.
- **Avoid Treating the Initiatives as Religion.** Organizations typically go through three phases during implementation. At first the initiative is

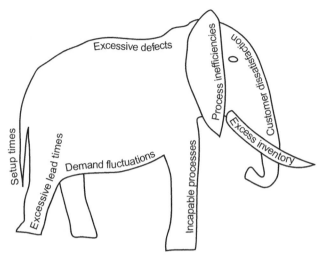

Figure C-1. The elephant of operational inefficiency.

seen as a set of tools to improve the operation. Then it is adopted as a philosophy that is used to guide all actions taken by a company. Sometimes people go a bit overboard and adopt the initiative as a "religion." People within companies become almost fanatical about the application. Although the enthusiasm can help with deployment, it can be detrimental for two reasons. First, the tools are applied in areas where another tool may be better suited. Second, the fanatical approach can turn off other employees.

No initiative is exempt from these problems (not even VRM). This chapter reviews several of the typical initiatives under way within various companies and describes how they relate to variation risk management. The initiatives reviewed in this chapter include Six Sigma, design for Six Sigma, Lean, variation reduction, dimensional management, design for manufacturability, House of Quality, and FMEA. This appendix only provides a quick summary: the reader should look to other texts and articles to fully understand the costs, strengths, and weakness of each of these initiatives.

C.1. SIX SIGMA

Six Sigma and Design for Six Sigma are currently in vogue for product and process improvement. They focus on significantly reducing the number of defects in processes. While the tools were initially applied to discrete manufacturing, they are now broadly applied to nearly all tasks executed as part of

The Parable of the Blind Men and the Elephant

It was six men of Indostan
To learning much inclined,
Who went to see the Elephant
Though all of them were blind,
That each by observation
Might satisfy his mind.

The First approached the Elephant
And, happening to fall
Against his broad and sturdy side,
At once began to bawl:
"God bless me, but the Elephant
Is very like a wall!"

The Second, feeling the tusk,
Cried, "Ho! what have we here
So very round and smooth and sharp?
To me 'tis very clear
This wonder of an Elephant
Is very like a spear!"

The Third approached the animal
And, happening to take
The squirming trunk within his hands,
Thus boldly up he spake:
"I see," quoth he, "The Elephant
Is very like a snake!"

. . .

And so these men of Indostan
Disputed loud and long,
Each in his own opinion
Exceeding stiff and strong.
Though each was partly in the right,
They all were in the wrong!

by John Godfrey Saxe (1816–1887)

an organization's activities including service businesses, organizational processes, and continuous processes. Six Sigma focuses on improving existing processes and design for Six Sigma focuses on developing new products that are robust to variation.

Variation risk management and the tools and processes described in earlier chapters can be used to augment and guide both initiatives. Variation risk management provides a set of tools to focus the tools and techniques on the critical areas. The tools for Six Sigma and design for Six Sigma tend to be general because they need to be as accessible to service organizations as to manufacturing organizations. The tools in this book provide methods that are specific to manufactured products, specifically discrete assembled products.

Six Sigma (Chowdhry, 2001; Pande et al., 2002a,b; Rath and Strong, 2002) had its start as an initiative at Motorola. It emphasizes controlling processes such that only 3.4 operations per million fall outside the acceptable tolerance range. The Six Sigma program, now implemented in a large number of companies, is based on existing using quality and variation reduction tools in a rigorous process and organizational context. Six Sigma emphasizes a methodical process: defining problems, measuring the current situation, analyzing the problem using statistical and quantitative methods, improving the process, and imposing ongoing control. Six Sigma also defines the organizational structure and support required to implement the program. For example, it involves creating a set of champions and experts (termed Green Belts and Black Belts).

While its tools and processes are effective, one of its main weaknesses is in the identification of cost-effective Six Sigma projects. In the program, teams are asked to look for opportunities for improvement and prove that these will have a significant ROI. However, Six Sigma implementation often overlooks systematic methods for examining an entire product delivery system and identifying the *best* opportunities for improvements. The use of systematic variation risk management can improve the effectiveness of a Six Sigma program.

Application of Six Sigma within an organization starts with developing the organizational structures to support it. This includes creating leadership councils, project sponsors, and implementation teams. Variation risk management can be used to justify to both the leadership councils and project sponsors that the best projects are being selected and that the potential improvements are real.

The I-A-M process also provides critical inputs to the Six Sigma process. During identification, a complete list of all possible areas for improvement is generated. During assessment, the major cost and quality drivers are identified and the weaknesses in the current quality management program are identified. The output of the assessment phase is a list of possible areas where Six Sigma projects may be applied. Once the projects are identified, teams can choose, based on feasibility, what areas to address. The define, measure, analyze, improve, and control (DMAIC) process that forms the backbone of the Six Sigma process can then be used to execute the improvements.

Details about each of the DMAIC steps can be found in a number of Six Sigma books (e.g., Chowdhry, 2002). The Six Sigma DMAIC process should start after the identification and assessment phases of VRM are completed. Requirements for the DMAIC process are partially fulfilled by work done in the identification and assessment phases. The integration of the two methods is shown in Figure C-2.

Define. This stage sets up the project definition, determines what customer requirements are being addressed, and performs preliminary process mapping. This phase also includes defining roles within the team, developing the project Gantt chart, and developing the return on investment analysis. The goal of this stage is to ensure that all parties, including management, understand the project, the resources required, and the potential outcomes.

The identification phase of the I-A-M procedures, which should be completed before selecting the projects, provides the project definition and customer requirements. Variation flowdown ensures that the projects worked on are related to issues of variability, not just poor design that did not meet the initial customer requirements. The assessment phase computes the total cost of variation and the relative contribution of each KC. This cost provides the measure of what the possible opportunity will be for cost savings.

Measure. This phase includes quantifying the measures of yield, defect, rate, causes, and so on. The data from this phase is used to identify where possible improvements may be implemented.

During the assessment phase of the I-A-M procedures, the major contributors and their costs are quantified. The assessment phase quantifies the opportunity but not necessarily the details as to why the contribution or cost is high. For example, the aggregated scrap costs may be available, but data may not be available on the individual causes for scrap. During the measurement phase the initial data analysis and data inventory that was done during assessment should be augmented with specific data collection on the problem being addressed.

Analyze. The difference between analyzing and measuring is important. Measuring defines *how* the manufacturing process performing and analyzing determines *why*. Once the manufacturing process is baselined and the major contributors are identified, it is up to the team to identify the root causes for the problems.

While the initial identification and assessment phases of the I-A-M procedures will provide the team with a picture of the total system, the analysis

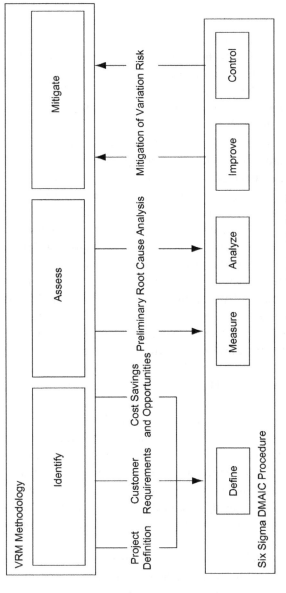

Figure C-2. Integrating VRM with Six Sigma.

phase of DMAIC gets into significant detail using design of experiments, root cause analysis tools, variation modeling, and process analysis tools to identify why the process is performing the way it is.

Improve. During this stage, ideas for improvements are generated and solutions are developed and deployed. It is important in this phase to choose the right solution, not the expedient one. Chapter 8 provides several methods for improving a process, including changing the design, changing the process, improving the process, controlling the process, and inspection/testing. The holistic view of variation and variation costs enables the team to select the best solution for the entire organization. There are many tools available in this phase. It is important to choose the right tool, not to simply choose the one you know best or try to apply them all.

Control. This phase answers which ongoing processes need to be monitored to ensure that the process continues to deliver quality product. The monitoring mitigation strategy as well as the section on transition to production and the optimal design of quality plans will help the team determine where to best place ongoing control.

When implementing Six Sigma in an organization, division, or plant, the leadership council, project sponsors, and implementation leaders should require a systems analysis of the product and/or processes to understand the best areas for improvement before jumping to select projects. Once the major opportunities have been prioritized, the list of possible projects can be generated.

There are several benefits to doing the homework of prioritization up front rather than once the projects are selected.

- The right projects are selected before anyone gets attached to any ideas. If someone comes up with an idea for a project, it is often difficult to get him or her to work on another person's ideas.
- A common and consistent method for measuring costs and benefits is predetermined. It is possible to compare apples to apples and not spend time understanding how the return on investment was determined.

Subsequent to project selection, systematic cost of variation analysis can act as a metric by which all Six Sigma projects are measured. By measuring the total cost of variation, the organization can track whether or not the improvements are impacting the total cost rather than "robbing Peter to pay Paul."

C.2. DESIGN FOR SIX SIGMA

Design for Six Sigma is an approach for applying Six Sigma principles to designing new products and services (Chowdhry, 2002; Tennant, 2002). The methods focus on ensuring that customer requirements are met. The typical DFSS method has several steps:

- **Identify and Define Opportunity.** This step looks at the customer needs and identifies opportunities to meet them in either a lower cost or better way.
- **Identify Requirements.** This step formalizes requirements using tools such as House of Quality.
- **Select Concepts.** This step generates and selects between concepts to achieve requirements.
- **Develop the Design.** This step generates the details of the design to deliver the product concept.
- **Optimize the Design.** This step involves applying robust engineering tools to make the product less sensitive to the existing variation.
- **Verify the Design.** This step validates that the design is capable of reliably delivering requirements.

The typical DFSS book presents general tools for each of these stages. The tools guide the team in identifying the correct customer requirements and determining how to achieve them. Most of the DFSS tools are good product development practices that have been available to the design community for many decades.

The benefit of the DFSS approach is its generality; it can be applied to services as well as manufactured products. However, because most product development and manufacturing organizations have a step-by-step product development process already in place, adopting DFSS in the form provided in the books can be difficult. Most companies implementing DFSS systems spend significant time designing a program that incorporates robust design into their existing product development process.

The information in Chap. 9 reviews the typical phases of product development; and how variation can be addressed in each stage. The principles of VRM and DFSS are consistent with each other and have a similar framework. The details in this book provide more information about how to design for variation in accordance with Six Sigma principles for the specific case of manufactured products.

C.3. LEAN MANUFACTURING

Lean is a term used (and misused) by many to describe the best practices in product development and production identified in the International Motor

Vehicle Program at MIT (Womack et al., 1990). That project identified practices that showed that Japanese companies were able to design and produce cars with shorter lead times, higher quality, and lower costs than were US companies. The collection of practices was termed the *Lean production system.* Since that time, Lean has been applied to service tasks, product development, and support tasks.

In general, Lean is concerned with the minimization of waste or *muda* (Murman et al., 2002; Womack and Jones 1996). This waste can come in the form of excess touch time, inventory, cycle time, capacity, rework, and scrap. Specifically for variation, it can include time wasted locating the root causes of problems, inspection, and redesign required when quality is not designed into a product.

The VRM methodology shares a similar philosophy. In the case of VRM, there are many opportunities to improve quality and reduce costs; however, some are better than others. Through data-driven methods, the VRM methodology works to aid in understanding where resources are best applied to reduce variation and its impact. VRM works to reduce waste in the most cost-effective manner. However, Lean addresses not only the impact of variation but also waste that is designed into the product and process.

Womack and Jones (1996) outline five practices that are critical to Lean. For each of the five practices, the related topics in VRM are discussed.

Specify Value. The organization must understand what customers want from the product and what the customer values. The VRM methodology is rooted in delivering customer requirements which are sensitive to variation. It is critical to the prioritization process as well as the Lean process to thoroughly understand what your customer values.

Identify the Value Stream. The value stream defines how the value to the customer is delivered through the entire organization. Identifying actions that do not support the creation of value enables possible waste to be identified. Lean places a strong emphasis on the flow of materials and information through the delivery of the product. When materials and/or equipment sit idle, capital is wasted and quality decreases. VRM accounts for value and waste by means of the total cost of defects.

Make Value Flow Continuously. This philosophy tries to avoid the batch-and-queue approach used in many production systems. It involves reducing inventory, reducing wait times, and reducing setup time. One of the more well known Lean practices is the reduction of inventory throughout the supply chain. Inventory is needed to buffer against many issues including differences in capacity, machine downtimes, scrap, and rework. Variation can have a significant impact on the need for inventory. The factory cannot risk having a bad batch of products arrive, requiring the line to be shut down. In addition, finished goods inventory may be needed if defects are caught at the end of the line and there may not be products available to ship to the customer.

Another typical Lean practice is the reduction of setup times. By reducing setup times, organizations can implement higher-mix, lower-volume production runs, reducing the need for inventory. Typically, setup times can be reduced by understanding the manufacturing process and having it under control. This can have a secondary benefit of decreasing excess variation often experienced in parts after a setup. By standardizing setups and streamlining them, the quality becomes more stable.

Value flow can also be enabled by regular maintenance of manufacturing equipment to avoid unexpected downtimes. Machine degradation, wear, and breakage can impact the quality of parts produced. Often a machine is brought down because the quality of parts produced falls below some acceptable level. When machines are proactively maintained, there is a lower chance that the features being produced will begin to drift away from nominal and toward the limit of acceptability.

By implementing owner inspection throughout the process puts quality in the hands of the person building the product. This reduces the number of faults that get built into the product but not caught until the end. In addition, it increases the chance that the attributable causes of variation will be caught by the operator and removed. Operator buy-off, along with single-piece flow, moves quality control as far upstream as possible, reducing the number of defects, the time to catch them, and the time to fix them.

Let Customers Pull Value. This approach uses *Kanban* systems in which the product is pulled through the factory rather than pushed. Kanban systems can only be implemented where there is minimal scrap or rework. They require each station to produce high-quality materials to ensure that a single-piece flow is possible. VRM focuses on the cost-effective attainment of the quality required to enable single piece flow.

Pursue Perfection. This strategy involves continuously improving products and processes to reduce waste. The improvement process is never finished; there is always a way to continue to streamline operations, take out redundant steps, improve quality, and ultimately remove waste. The continual improvement should be based on data-driven and robust methods for root cause analysis and repair. Variation risk management in production is one of the multiple efforts that should be in place to reduce cost and improve quality. Continually highlighting the most costly areas and addressing them allows the cost of the product to continue to drop while quality is maintained or improved.

C.4. CONTINUAL IMPROVEMENT, TQM, AND KAIZEN

Continual improvement, TQM, and Kaizen all have the goal of making the organization and its products better. Both TQM and continual improvement are typically focused on organization improvement and have very broad goals, only one of which is variation reduction.

Total quality management is a collected set of tools that encourages the entire organization to use quality thinking and tools to manage all aspects of an organization. TQM uses a broad definition of quality that focuses on the general satisfaction of the customer and includes delivering the right product with short product development cycles as well as reliably delivering the product with low variation. It, like other initiatives, is based on existing tools such as house of quality, cause and effect diagrams, and SPC. Many of the TQM programs also include strategic plans and organizational change tools to adjust the behavior of the entire organization.

Continual improvement is a set of tools to focus an organization on improving aspects of an organization such as quality, business results, and customer satisfaction. It is a systematic approach to setting goals for an organization, creating metrics, and tracking the metrics as they improve. It uses tools such as balanced score cards to track target goals and the achievement toward them.

One specific tool used to achieve continual improvement is Kaizen, whose name comes from the Japanese word for continual improvement. The goal of Kaizen is to challenge the people doing work to find better ways of doing their jobs. The Kaizen methods work to improve processes without major capital expenditures. Typically these methods are applied in very focused improvement efforts that work to remove waste involved in setup, wait time for products, inventory, finding equipment, moving between locations, and so on. The focused improvement efforts tend to be short events (several days) that result in dramatic changes to the way business is done. The Kaizen events are typically not based on rigorous and in-depth data analysis.

Kaizen can have some side benefits in the reduction of variation through standard operating procedures, less touch labor, lower error rates, and so on. However, typically Kaizen events are not focused specifically on issues related to variation and quality.

C.5. DIMENSIONAL MANAGEMENT

Dimensional management is a subset of variation risk management and is specific to geometric issues (Liggett, 1993). Dimensional management tools include procedures and methods used to set, analyze, and monitor tolerances. In a broader scope, it defines how the dimensions of a product are achieved through appropriate product decompositions, assembly methods, locating features, and datum structures—all of which drive how variation will impact the final dimensional targets of the product. Dimensional management focuses on creating products with common indexing and datuming schemes.

Dimensional management is especially important where tooling and fixture schemes must be determined early because of the lead time on tooling and because the tooling and fixtures are expensive to change. The same VRM framework of I-A-M is used for dimensional management.

C.6. DESIGN FOR MANUFACTURING

The broad class of design for manufacturing tools (Bralla, 1998; Corbett et al., 1991; Dixon and Poli, 1999) have the general goal of encouraging the creation of designs that are more manufacturable. The primary goal of DFM tools is to reduce the recurring fixed costs including parts, material, and labor. In general, DFM tools are made up of heuristic rules that should be followed by the designer to ensure better manufacturability. Some of these rules can help reduce the impact of variability; however, the rules are not exclusively related to variation. Typical DFM tools include:

- **Design for Assembly (DFA).** These tools are used to improve assembly-specific issues, including assembly time, errors, material costs, and so on (Dewhurst et al., 2001) The rules for DFA (i.e., designing parts with clear orientations) can reduce the chances of assembly errors and can reduce the chance of part damage during assembly.
- **Process-Specific Guidelines.** These guidelines are specific to individual processes such as machining or surface-mount printed circuit board. For example, the DFM guidelines may indicate what kind of features can and cannot be created by a process (e.g., blind pockets in a two-degree-of-freedom end mill) or what features are difficult to create and can cause excess scrap (e.g., oversized parts requiring hand placement on a surface-mount printed circuit board).

Variation risk management is more than just a DFM tool. It focuses the organization on where DFM may be needed to address variation issues. In addition, there may be areas where DFM is used to reduce the recurring costs of producing a product where variation has no or only an indirect impact (i.e., reducing assembly time, processing time, or part count).

C.7. QUALITY FUNCTION DEPLOYMENT (HOUSE OF QUALITY)

Quality function deployment (QFD) also called the House of Quality is used to systematically translate the voice of the customer into clear technical requirements. It is used to ensure that customer needs are met. In addition, it provides a method for comparing the target requirements to those of the competitors. Typically it is used in the requirements definition phase of product development.

Figure C-3 shows the structure of a typical House of Quality. The main part of the tool is the matrix that relates the voice of the customer to the technical requirements. Several books are available on the subject that detail how the House of Quality is used (Cohen, 1995; Revelle et al., 1998)

The extended House of Quality (Figure C-4) attempts to continue the process from the voice of the customer through to the process control parameters. For each level of decomposition, a new house is developed. The final house can be used to compare the process specifications to the current capability. There are

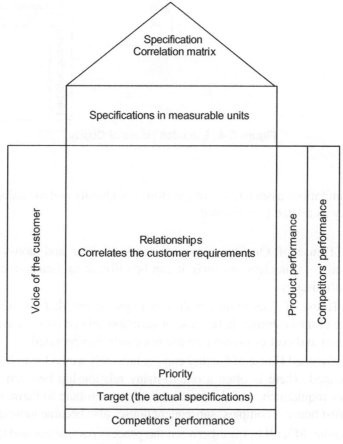

Figure C-3. House of Quality.

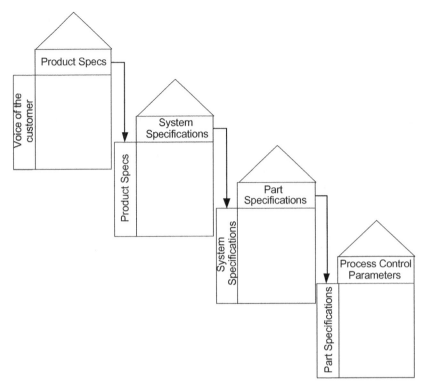

Figure C-4. Extended House of Quality.

several limitations associated with the House of Quality and extended House of Quality as related to variation:

- The House of Quality mixes variation-sensitive and non-variation-sensitive parameters. As such, it can be difficult to focus on variation-specific issues.
- The House of Quality only addresses requirements that result from the voice of the customer. In the case of variation-related issues, ease of production and cost of production are not easily incorporated.
- The extended House of Quality can get unwieldy as the lower houses are generated. There is often a one-to-many relationship between the customer requirement and the parameters that contribute to them. Often, if the first house is complex, subsequent houses also become more complex.
- It can be difficult to trace between the process parameters and the voice of the customer. The representation can be awkward to maintain and check for errors.

Where systems have requirements that are tightly coupled and where there are few layers between the product, part, and process KCs, the house of quality representation can highlight key trade-offs and can provide a compact representation scheme for variation flowdown.

C.8. FMEA

Failure modes and effects analysis is a tool used to qualitatively assess possible failure modes and identify possible causes (McDermott et al., 1996). FMEA is used in a number of industries for proactively identifying potential problems in the concept, product or production system. In-house and customers' environmental health, and safety issues are often included. FMEA is limited in its ability to handle complex flowdowns especially where a single cause can impact multiple failure modes.

The FMEA process starts by brainstorming all possible failure modes and then assigning each one a rating of 1 to 10 on severity, occurrence, and detection. Severity measures how detrimental the impact of the failure would be on the customer. Occurrence measures how often the failure is likely to happen. Detection measures whether or not the current quality control system is capable of detecting the failure before it impacts the customer. The ranking is then used to prioritize what failures should be addressed first by teams.

There are several limitations when applying FMEA to variation-sensitive issues:

- The flowdown from a system failure to part dimension can be difficult to document and trace. In addition, critical processes that contribute to multiple failure modes cannot be easily identified.
- The severity/occurrence/detection rating combines a measure of the design capability and the mitigation capability. If a critical issue is expensive and has a high chance of occurring, but the current quality control is easily capable of identifying it, the issue appear lower in the list. This does not encourage teams to design out either the severity or the occurrence.
- The FMEA table does not separate variation-specific from other design related issues.

C.9. SUMMARY

The quality improvement world is full of tools, including Kaizen, Design for Six Sigma, Six Sigma, Lean, and TQM. This appendix briefly reviewed a

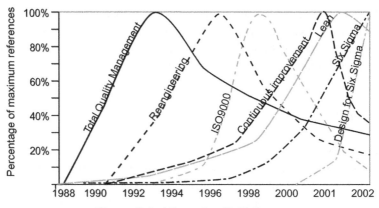

Figure C-5. Life cycle of initiatives.

small subset of them. Most of the tools work to minimize waste through intelligent decision making, concentration on improvements, and systems thinking and are focused on improving the operational efficiency of an organization. They should be viewed as a varied toolbox from which the organization can pull to address very specific problems. New tools are often merely repackaging of existing old tools and philosophies with new buzz words and slogans. Figure C-5 shows the natural life cycle of various initiatives. The graph was generated by counting the number of times articles in the general press contained references to a given method. The graph is normalized to show, for each year, the number of publications relative to the maximum number of citations in any year. The graph shows that Lean, Six Sigma, and design for Six Sigma still have "life," but, given that history repeats itself, their curves are likely to follow the same path as those of the other initiatives. Unfortunately, the attention span of organizations often requires the newest fad to keep attention on quality. Hopefully the philosophy of quality, improvement, data-driven methods, and focusing on variation will continue on in organization long after the jargon is gone.

APPENDIX D

SUMMARY OF PROCESS DIAGRAMS

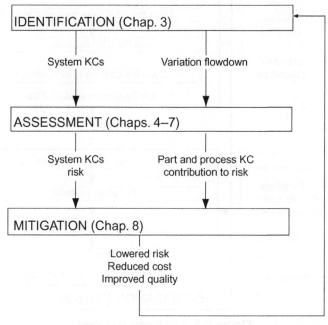

Figure D-1. Overview of the identification, assessment, and mitigation process.

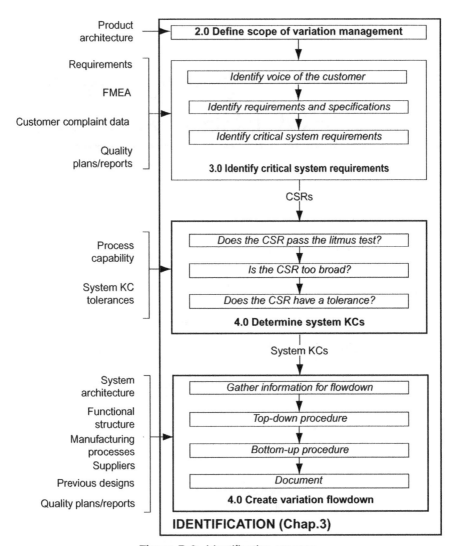

Figure D-2. Identification process.

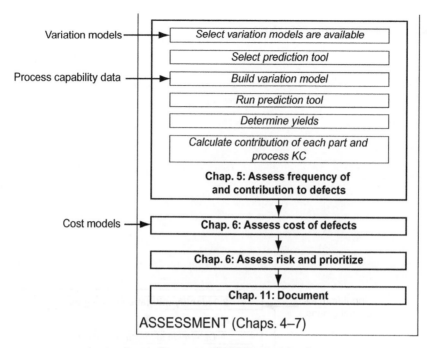

During product development

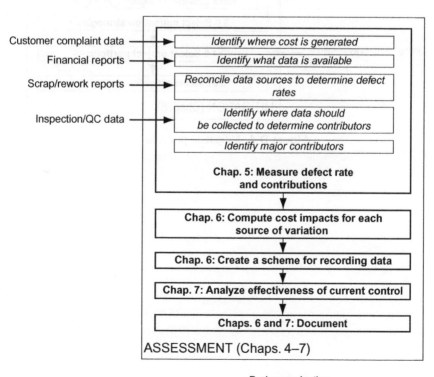

During production

Figure D-3. Assessment process.

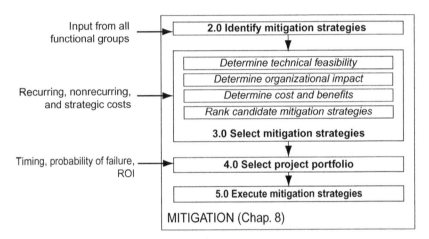

Figure D-4. Mitigation process.

GLOSSARY

Assessment. The process of quantifying the total impact of variation on the safety, perception, performance, and end cost of the product.

Attributable causes of variation (special causes of variation). One-time events that cause a process to go out of control (vary more than expected) and that can be attributed to a specific action, inaction, or event.

Bottom-up variation flowdown process. The process of creating the variation flowdown starting with the part and process critical dimensions or features (part and process KCs) and linking them to the system KCs.

Capability. The ability of a process to reliably achieve the target tolerances, usually expressed as C_{pk}. The capability is a function of both the mean and the standard deviation. The short-term capability is the capability measured when the process is in control and over a single shift. The long-term capability includes the changes in capability over time.

Capacity. The number of parts or products that can be produced by a production line in a unit of time.

Characteristic (dimension or feature). Terms used for any number of product characteristics including measured dimensions, functional aspects of a product such as forces and voltages, shape, surface finish, and materials.

Contribution. The relative amounts that individual part or process KCs impact the defect rate of a system KC.

Critical System Requirements A subset of the most important requirement for a system. They can include performance, manufacturability and interface requirements.

Cycle time. The elapsed time required to produce a single part or product. Cycle time is the inverse of throughput (the total number of parts or products produced in a set time period).

Defect. An occurrence of a system characteristic falling outside its allowable tolerance.

Dimension. See Characteristic.

Dimensional management. The tools used to manage and design for the variation in individual dimensions. These can include GD&T, datum schemes, and measurement methods.

Expected cost. See risk.

External noise factor. The sources of variation outside the control of the design and/or production, including vibration, noise, wear, environment, and use.

Feature. See Characteristic

Gate. See Stage gate.

Holistic view of variation. Looking at the impact of variation in parts and processes and its effect on the total quality of a product as well as at the cost, impact, and capability of all functional groups that influence or are influenced by variation.

Identification. The process of finding all dimensions or features that are sensitive to variation and that might impact the customer perception, performance, or end cost of the product.

Key characteristic. A feature of a system, assembly, part, or process whose expected variation from nominal has an unacceptable impact on cost, performance, or safety.

Latitude. See Tolerance

Margin. See Tolerance

Mitigation strategy. The approach taken to reduce variation or the impact of variation.

Nominal value. See target value.

Process capability database. A database containing the capability of existing manufacturing to deliver the tolerances on part and process features.

Product development process. The complete process of developing, designing, and producing a product for delivery to a customer.

Production. The steps to produce a product in the manufacturing environment including all tasks in the manufacturing process.

Quality control. Efforts to prevent defects from being produced or from escaping from the factory to the customer. The efforts can take the form of statistical process control, inspection, testing, or sampling.

Quality control effectiveness. How well the current quality control plan prevents defects from reaching the customer.

Quality control maturity. How well the current quality control plan prevents defects from reaching the customer, allows for rapid identification of root causes, and efficiently uses limited QC resources.

Risk (Expected cost). The cost of a defect times the probability of that defect.

Robust design. An approach to design that makes a product less sensitive to the variation introduced by manufacturing, assembly, users, or the environment.

Special causes of variation. See Attributable causes of variation.

Specification. A combination of a target value and tolerance for a characteristic.

Stage. See stage gate.

Stage gate. An approach to product development that uses preplanned reviews conducted at the end of each stage of the product development process.

Station. A point in the manufacturing process where value is added to the product (work or assembly) or the product is inspected and/or reworked.

Statistical process control. A charting method used to track the behavior of a system over time to identify where processes are going out of control because of attributable causes of variation.

System. A product or major part of a product, typically the design or production responsibility of a single IPT. A system is composed of assemblies and parts.

Target value (nominal value). The desired or specified value for a feature or dimension. The target value is typically the value indicated on the specification sheet and/or drawing.

Tolerance (latitude, margin). The maximum amount of variation allowed in a feature, dimension, or specification.

Tolerance allocation. The process of allocating the allowable variation in the customer requirements to the individual parts and processes.

Tolerance analysis. The process of rolling up the capability of individual parts and processes to determine the impact of variation on customer requirements.

Tolerance assessment. The process of evaluating whether or not individual tolerances can be achieved with the existing process capability. Evaluation can be done either through manufacturing reviews or by using process capability databases.

Top-down variation flowdown process. The process of creating the variation flowdown, starting with the system KCs and methodically identifying the KCs of the assemblies, parts, and processes.

Variation flowdown. A diagram that provides a systems view of variation risk factors by portraying the hierarchy of variation-sensitive critical system requirements and the part and process features that contribute to the variation of the CSRs.

Variation risk management. The systematic allocation, based on cost and risk, of limited resources to reduce variation or the impact of variation on cost, performance or safety.

Variation reduction. The reduction of variation introduced by manufacturing, environmental, and/or assembly processes.

BIBLIOGRAPHY

Andersen, B. (ed.). (1999). *Root Cause Analysis: Simplified Tools and Techniques.* Milwaukee, WI, American Society of Quality.

Bardwick, J. M. (1995). *Danger in the Comfort Zone: From Boardroom to Mailroom—How to Break the Entitlement Habit that's Killing American Business.* New York, Amacon.

Box, G.E.P., J. S. Hunter, et al. (1978). *Statistics for Experimenters: An Introduction to Design, Data Analysis, and Model Building.* New York, John Wiley & Sons.

Boyer, D. E. and J. W. Nazemetz. (1985). "Introducing Statistical Selective Assembly—A Means of Producing High Precision Assemblies from Low Precision Components." Annual International Industrial Engineering Conference.

Bralla, J. G. (1998). *Design for Manufacturability Handbook.* New York, McGraw-Hill Professional.

Chowdhry, S. (2001). *The Power of Six Sigma: An Inspiring Tale of How Six Sigma Is Transforming the Way We Work.* Dearborn, MI, Dearborn Trade.

Chowdhry, S. (2002). *Design for Six Sigma: The Revolutionary Process for Achieving Extraordinary Profits.* Dearborn, MI, Dearborn Trade.

Cleland, D. I. (1996). *Strategic Management of Teams.* New York, John Wiley & Sons.

Cohen, L. (1995). *Quality Function Deployment.* Englewood Cliffs, New Jersey, Prentice Hall PTR.

Corbett, J., M. Dooner, et al. (1991). *Design for Manufacture: Strategies, Principles and Techniques.* Reading, MA, Addison-Wesley.

Crosby, P. B. (1979). *Quality Is Free: The Art of Making Quality Certain.* New York, McGraw-Hill Professional.

Deming, W. E. (1986). *Out of the Crisis: Quality, Productivity and Competitive Position.* Cambridge, UK, Cambridge University Press.

Dewhurst, P., W. Knight, et al. (2001). *Product Design for Manufacture & Assembly Revised & Expanded.* New York, Marcel Dekker.

Dickinson, M. W., A. C. Thornton, et al. (2001). "Technology Portfolio Management: Optimizing Interdependent Projects over Multiple Time Periods." *IEEE Transactions on Engineering Management* 48(4):518–527.

Dixon, J. and C. Poli (1999). *Engineering Design & Design for Manufacturing: A Structured Approach.* Conway, MA, Field Stone.

Fisher, K. (1999). *Leading Self-Directed Work Teams.* New York, McGraw-Hill.

Fowlkes, W. Y. and C. M. Creveling (1995). *Engineering Methods for Robust Product Design: Using Taguchi Methods in Technology and Product Development.* Reading, MA, Addison-Wesley.

Fulgum, D. A. (1994a). "F-22 Fix to Cost $ 20–25 Million." *Aviation Week & Space Technology* 140(20):27.

Fulgum, D. A. (1994b). "F-22 Signature Problem Inflicts Weight Penalty." *Aviation Week & Space Technology* 140(11):30.

Galloway, D. (1994). *Mapping Work Processes.* Milwaukee, WI, American Society for Quality.

Gano, D. L. (1999). *Apollo Root Cause Analysis—A New Way Of Thinking.* Yakima, WA, Apollonian Publications.

Gross, J. M. (2002). *Fundamentals of Preventive Maintenance.* New York, Amacom.

Hanson, J. (2001). *Improving Process Capability Data Access for Design.* Boston, MIT.

Hauser, J. R. and D. Clausing (1988). "The House of Quality." *Harvard Business Review.* 66(3):63–74.

Hopp, W. J. and M. L. Spearman (2000) *Factory Physics.* New York, McGraw-Hill.

Juran, J. M. and A. B. Godfrey (eds.) (1998). *Juran's Quality Handbook.* New York, McGraw-Hill.

Kern, D. (2003). *Forecasting Manufacturing Variation Using Historical Process Capability Data: Applications for Random Assembly, Selective Assembly and Serial Processing.* Ph.D. thesis. Cambridge, MA, MIT.

Leyland, C. (1997). *A Cultural Analysis of Key Characteristic Section and Team Problem Solving during an Automobile Launch.* Boston, M.I.T.

Liggett, J. V. (1993). *Dimensional Variation Management Handbook: A Guide for Quality, Design, and Manufacturing Engineers.* Englewood Cliffs, NJ, Prentice Hall.

McDermott, R. E., R. J. Mikulak, et al. (1996). *The Basics of FMEA.* Portland, OR, Productivity Inc.

McDonnell Douglas (1997). "First New Landing Gear Pads Installed on C-17 Globemaster III." Press Release #97-30.

McGrath, M. E. (ed.) (1996). *Setting the PACE in Product Development.* Boston, Butterworth-Heinemann.

Meyer, M. H. and A. P. Lehnerd (1997). *The Power of Product Platforms: Building Value and Cost Leadership.* New York, The Free Press.

Montgomery, D. C. (1996). *Introduction to Statistical Quality Control.* New York, John Wiley & Sons.

Montgomery, D. C. (2000). *Design and Analysis of Experiments.* New York, John Wiley & Sons.

Murman, E., T. Allen, et al. (2002). *Lean Enterprise Value: Insights from MIT's Lean Aerospace Initiative.* New York, Palgrave.

Nagler, G. (1996). *Sustaining Competitive Advantage in Product Development: A DFM Tool for Printed Circuit Assembly.* M.S. thesis. Cambridge, MA, MIT.

National Transportation Safety Board (NTSB). (2002). *Aircraft Accident Report, Loss of Control and Impact with Pacific Ocean Alaska Airlines Flight 261 Mc-Donnell Douglas MD-83, N963AS about 2.7 Miles North of Anacapa Island, California, January 31, 2000.* NTSB/AAR-02/01 PB2002-910402, December 30, 2002.

Pande, P. S., R. P. Neuman, et al. (2002a). *The Six Sigma Way: How GE, Motorola, and Other Top Companies are Honing their Performance by Honing their Performance.* New York, McGraw-Hill Professional.

Pande, P. S., R. P. Neuman, et al. (2002b). *The Six Sigma Way Team Fieldbook.* New York, McGraw-Hill.

Phadke, M. S. (1989). *Quality Engineering Using Robust Design.* Englewood Cliffs, NJ, Prentice Hall PTR.

Rath & Strong, Inc. (2002). *Rath & Strong's Six Sigma Pocket Guide.* Lexington, MA, Rath & Strong, Inc.

Revelle, J. B., J. W. Moran, et al. (1998). *The QFD Handbook.* New York, John Wiley & Sons.

Ross, P. J. (1996). *Taguchi Techniques for Quality Engineering: Loss Function, Orthogonal Experiments, Parameter and Tolerance Design.* New York, McGraw-Hill.

Shimbun, N. K. (ed.) (1989). *Poka-Yoke: Improving Product Quality by Preventing Defects.* Portland, OR, Productivity Press.

Shingo, S. (1986). *Zero Quality Control: Source Inspection and the Poka-Yoke System.* Portland, OR, Productivity Press.

Society of Automotive Engineers (SAE). (2001). *Variation Management of Key Characteristics,* SAE AS9103.

Taguchi, G. (1992). *Taguchi on Robust Technology Development: Bringing Quality Engineering Upstream.* New York, ASME Press.

Tata, M. and A. Thornton (1999). "Process capability usage in industry: Myth vs. Reality." Design for Manufacturing Conference, ASME Design Technical Conferences, Las Vegas, NV. DETC 99/DFM-8968.

Tennant, G. (2002). *Design for Six Sigma: Launching New Products and Services without Failure.* Burlington VT, Gower.

Thornton, A. C. (1999). "A Mathematical Framework for the Key Characteristic Process." *Research in Engineering Design* 11(3):145–157.

Thornton, A. C. (2001a). "Optimism vs. Pessimism: Design Decisions in the Face of Process Capability Uncertainty." *ASME Journal of Mechanical Design* 123(3):313–321.

Thornton, A. (2001b). "The Use of Process Capability Data in Design." In *Data Mining for Design and Manufacturing: Methods and Applications,* D. Braha (ed.). Dordrecht, Netherlands, Kluwer Academic Publishers, pp. 505–518.

Thornton, A. C., S. Donnelly, et al. (2000). "More than Just Robust Design: Why Product Development Organizations Still Contend with Variation and Its Impact on Quality." *Research in Engineering Design* 12(3):127–143.

Thornton, A. C. and M. Tata (2000). "Use of Graphical Displays of Process Capability Data to Facilitate Producibility Analyses." *Artificial Intelligence for Engineering Design, Analysis and Manufacturing* 14(3):181–192.

Ulrich, K. T. and S. D. Eppinger (1995). *Product Design and Development.* New York, McGraw-Hill.

Wheelwright, S. C. and K. B. Clark (1992). *Revolutionizing Product Development: Quantum Leaps in Speed, Efficiency, and Quality.* New York, Free Press.

Womack, J. P. and D. T. Jones (1996). *Lean Thinking.* New York, Simon & Schuster.

Womack, J. P. and D. T. Jones (2002). *Seeing the Whole: Mapping the Extended Value Stream.* Brookline, MA, Lean Enterprises Institute.

Womack, J. P., D. T. Jones, et al. (1990). *The Machine that Changed the World.* New York, Rawson Associates.

INDEX

Printed and bound by CPI Group (UK) Ltd, Croydon, CR0 4YY

23/04/2025

14660924-0005